M000204369

Sustainable Urban Agriculture in Cuba

Contemporary Cuba

UNIVERSITY PRESS OF FLORIDA

Florida A&M University, Tallahassee
Florida Atlantic University, Boca Raton
Florida Gulf Coast University, Ft. Myers
Florida International University, Miami
Florida State University, Tallahassee
New College of Florida, Sarasota
University of Central Florida, Orlando
University of Florida, Gainesville
University of North Florida, Jacksonville
University of South Florida, Tampa
University of West Florida, Pensacola

Sustainable Urban Agriculture in Cuba

SINAN KOONT

University Press of Florida

Gainesville · Tallahassee · Tampa · Boca Raton
Pensacola · Orlando · Miami · Jacksonville · Ft. Myers · Sarasota

22 21 20 19 18 17   6 5 4 3 2 1

First cloth printing, 2011
First paperback printing, 2017

Library of Congress Cataloging-in-Publication Data
Koont, Sinan, 1943-
Sustainable urban agriculture in Cuba / Sinan Koont.
p. cm.—(Contemporary Cuba)
Includes bibliographical references and index.
ISBN 978-0-8130-3757-8 (cloth: alk. paper)
ISBN 978-0-8130-5403-2 (pbk.)
1. Sustainable agriculture—Cuba. 2. Urban agriculture—Cuba.
3. Agriculture and state—Cuba. I. Title. II. Series: Contemporary Cuba.
S477.C9K66 2011
630.97291'091732—dc23
2011028137

The University Press of Florida is the scholarly publishing agency for the
State University System of Florida, comprising Florida A&M University,
Florida Atlantic University, Florida Gulf Coast University, Florida
International University, Florida State University, New College of Florida,
University of Central Florida, University of Florida, University of North
Florida, University of South Florida, and University of West Florida.

University Press of Florida
15 Northwest 15th Street
Gainesville, FL 32611-2079
http://www.upf.com

Contents

Figures

Tables

Acknowledgments

First of all, I would like to express my gratitude for the support and understanding I received from members of my family, especially my spouse, Madelyn Campbell, during the two years it took me to complete this book. In fact, some members of my family turned out to be careful and prolific editorial assistants and deserve my heartfelt thanks: my sisters, Sevin Koont and Esin Suna Kunt; my son Shaun Koont; and my brother-in-law, Douglas Campbell.

In Cuba my work was aided by the Facultad Latinoamericana de Ciencias Sociales (FLACSO) at the University of Havana, which hosted my sabbatical stay in 2007. I could not have accomplished anything without the support of Beatriz Díaz González, then director of FLACSO and a longtime friend and colleague since my initial stay in Cuba in 1994. At FLACSO thanks are also due to Reynaldo Jiménez, who shared his office and Internet access with me, and to Dr. Mireya Sanz Medina, a scholar of urban agriculture in her own right, who was tireless in her efforts to arrange interviews and access for me. Three recent graduates of FLACSO, Gema González Hernández, Adele Cuba, and Elvira O'Reilly Morris, who had written master's theses on various aspects of urban agriculture in Ciudad de la Habana, consented to be interviewed by and to discuss their work with me. Elvira O'Reilly Morris also arranged for me to visit the agricultural cooperative where she now works.

Dr. Nelso Companioni Concepción, executive secretary of Grupo Nacional de Agricultura Urbana, and Eugenio Fuster Chepe, president of Asociación Cubana de Técnicos Agrícolas y Forestales, both prominent leaders in the urban agriculture movement in Cuba, took time from their busy schedules to give me multiple and lengthy interviews and arrange my visits to urban agriculture sites. Others who gave of their time, discussed their roles in the urban agriculture of Havana, and shared their insights with me include Miguel Salcines López, the leader at an urban agricultural cooperative in Alamar, Havana, and one of the most impressive overachievers of Cuban urban agriculture, an intelligent, humble, and dedicated man; and Dr. Raúl Gil Sánchez of La Hierba Buena, a prominent figure among Cuban mental health professionals and an urban agriculturalist par excellence. Their willingness to help with my studies was indispensable to this book.

In the United States, Dickinson College, the institution where I teach, made it all possible through its support of my efforts to study urban agriculture in Cuba. It granted me the sabbatical leave to begin with, and once a book project materialized it gave me financial summer support both directly and in the form of a very capable student editorial assistant, Alexander Bloom, class of 2011. I also obtained a reduction in my teaching load so that I could continue working on the book during the academic year.

And finally, I owe a debt of gratitude to Dr. John Kirk of Dalhousie University, the academic editor of the University Press of Florida series on Cuba, who encouraged me greatly, both before and during the writing of the manuscript. I also received kind and helpful editorial guidance from Amy Gorelick, the acquisitions editor at the University Press of Florida, and from Kirsteen E. Anderson, who played important roles in getting the manuscript into its final shape.

Abbreviations

ACC	Academia de Ciencias Cubanas (Cuban Academy of Sciences)
ACPA	Asociación Cubana de Producción Animal (Cuban Association of Animal Production)
ACTAF	Asociación Cubana de Técnicos Agrícolas y Forestales (Cuban Association of Agricultural and Forestry Technicians)
ANAP	Asociación Nacional de Agricultores Pequeños (National Association of Small Farmers)
ANIR	Asociación Nacional de Innovadores y Racionalizadores (National Association of Innovators and Rationalizers)
BTJ	Brigada Técnica Juvenil (Youth Technical Brigade)
CCS, CCSF	Cooperativa de Créditos y Servicios (Fortalecida) (Cooperative for Credits and Services [Strengthened])
CDR	Comité de Defensa de la Revolución (Committee for Defense of the Revolution)
CEAS	Centro de Estudios de Agricultura Sostenible (Center for Sustainable Agriculture Studies)
CENIC	Centro Nacional de Investigaciones Científicas (National Center for Scientific Research)
CENSA	Centro Nacional de Sanidad Agropecuaria (National Center for Plant and Animal Health)
CETAS	Centro de Estudios para la Transformación Agraria Sostenible (Center of Studies for Sustainable Agrarian Transformation)

CIGB	Centro de Ingeniería Genética y Biotecnología (Genetic Engineering and Biotechnology Center)
CIM	Centro de Inmunología Molecular (Molecular Immunology Center)
CITMA	Ministerio de Ciencia, Tecnología y Medio Ambiente (Ministry of Science, Technology, and the Environment)
COMECON	Council for Mutual Economic Assistance
CPA	Cooperativa de Producción Agropecuaria (Agricultural Production Cooperative)
CREEs	Centros de Reproducción de Entemopatógenos y Entomófagos (Centers for the Reproduction of Entemopathogens and Entemophages)
CTA	*consultorio-tienda del agricultor* (consultancy and store for the agriculturalist)
CUC	*peso convertible* (convertible Cuban peso)
CUP	Cuban peso (also MN, meaning *moneda nacional,* or national currency)
EJT	Ejército Juvenil de Trabajo (Work Army of the Youth)
ENS	Empresa Nacional de Semillas (National Seed Company)
FAO	Food and Agriculture Organization of the United Nations
FCT	Fórum de Ciencia y Técnica (Forum of Science and Technology)
FLACSO	Facultad Latinoamericana de Ciencias Sociales (Latin American College of Social Sciences)
FMC	Federación de Mujeres Cubanas (Federation of Cuban Women)
GDP	gross domestic product
GMAU	Grupo Municipal de Agricultura Urbana (Municipal Group for Urban Agriculture)
GNAU	Grupo Nacional de Agricultura Urbana (National Group for Urban Agriculture)

GPAU	Grupo Provincial de Agricultura Urbana (Provincial Group for Urban Agriculture)
HORTIFAR	Armed Forces Horticultural Enterprise
IBP	Instituto de Biotecnología de las Plantas (Institute of Plant Biotechnology)
ICA	Instituto de Ciencia Animal (Institute of Animal Science)
IDRC	International Development Research Centre (Canada)
IIHLD	Instituto de Investigaciones Hortícolas "Liliana Dimitrova" (Liliana Dimitrova Institute for Horticultural Research)
IIRD	Instituto de Investigaciones de Riego y Drenaje (Institute of Research on Irrigation and Drainage)
INCA	Instituto Nacional de Ciencias Agrícolas (National Institute of Agricultural Sciences)
INIFAT	Instituto de Investigaciones Fundamentales en Agricultura Tropical "Alejandro de Humboldt" (Alexander Humboldt Institute of Fundamental Research in Tropical Agriculture)
INISAV	Instituto de Investigaciones en Sanidad Vegetal (Institute for Plant Health Research)
INIVIT	Instituto Nacional de Investigaciones de Viandas Tropicales (National Research Institute for Tropical Crops)
INRE	Instituto Nacional de Reserva Estatal (National Institute of State Reserves)
IPA	Instituto Politécnico Agropecuario (agricultural vocational school)
IS	Instituto de Suelos (Institute of Soils)
LER	land-equivalent ratio
MAE	*mercado agropecuario estatal* (state market for agricultural goods)
MAL	*mercado agropecuario libre* (free market for agricultural goods)

MES	Ministerio de Educación Superior (Ministry of Higher Education)
MINAG	Ministerio de la Agricultura (Ministry of Agriculture)
MINAZ	Ministerio de Azúcar (Ministry of Sugar)
MINED	Ministerio de Educación (Ministry of Education)
MINFAR	Ministerio de las Fuerzas Armadas Revolucionarías (Ministry of Revolutionary Armed Forces)
MININT	Ministerio del Interior (Ministry of the Interior)
MINSAP	Ministerio de Salud Pública (Ministry of Public Health)
PE	*perfeccionamiento empresarial* (improvement/enhancement of management)
SANE	Sustainable Agriculture Networking and Extension
SCIT	Sistema de Ciencia y Innovación Tecnológica (System of Science and Technological Innovation)
SICS	Sistema de Inspección y Certificación de Semillas (Inspection and Certification System for Seeds)
TRD	*tiendas de recaudación de divisas* (foreign currency collection stores, popularly referred to as "dollar stores")
UBPC	Unidad Básica de Producción Cooperativa (Basic Unit of Cooperative Production)
UCLV	Universidad Central de Las Villas (Central University of Las Villas)
UJC	Unión de Jóvenes Comunistas (Union of Young Communists)
UNAH	Universidad Agraria de la Habana "Fructosa Rodríguez Pérez" (Fructosa Rodríguez Pérez Agrarian University of Havana)
UNDP	United Nations Development Programme
UN-HABITAT	United Nations Human Settlement Programme

Introduction

When Cuba lost its capacity to import goods following the collapse of trade relations with the Soviet Union and the Soviet-led Council for Mutual Economic Assistance (COMECON) trading network in the early 1990s, it faced a crisis of survival. Within the space of a few years, the highly industrialized agriculture it had historically practiced was no longer feasible for lack of imported petroleum, fertilizers, pesticides, and agricultural machinery. In response the country undertook a dramatic and unprecedented reorganization of its agricultural system, turning to the agroecological production of food in or near cities and to a focus on growing food for local consumption rather than crops for export. The government's ultimate goal is to make the island self-sufficient in meeting the nutritional needs of the population through small-scale agriculture reliant on human labor rather than machinery and chemicals. In this effort, Cuba has achieved considerable success, especially in the production of fresh vegetables. As chapters 8 and 9 document, Cuba's success has been uneven; the surprising level of success in raising fruits and vegetables is balanced against much more modest progress in raising livestock for protein and in restructuring rural agriculture along more sustainable lines. Also, the extent to which the Cuban government and populace have a philosophical commitment to urban agriculture that will survive the easing of the economic crisis remains unclear. This book aims not only to report on the successes and challenges of this experiment in urban agriculture but also to analyze its underpinnings in historical roots, current policies, and institutional innovations.

The most important historical root undoubtedly is the Cuban government's consistent efforts directed at human resource formation since 1959: universal education at the primary and secondary levels, a rapidly expanding system of higher education, an emphasis on research and development in universities and in specialized research centers independent of universities, innovation in technology, and diffusion of that technology through training networks. These comprehensive investments in science, technology, and education, including in the area of agricultural sciences, were made over the 30 years between the Cuban Revolution and the implosion of the Soviet Union. This infrastructure ensured that when the crisis came, Cuba possessed a very high level of knowledge in science and technology, particularly for a third-world country. The country could also count on the availability of a universally literate, generally skilled, and educated workforce—adaptable, capable of learning new skills and engaging in research aimed at turning development efforts in a new direction.

Another significant historical legacy was the existence in revolutionary Cuba of a strong and coherent gender-, work-, and territorially based social organization: labor unions, the Federation of Cuban Women, student organizations, neighborhood committees, and small farmers associations. These social networks facilitated the government's task of engaging in organizational innovation and implementing policy concerning urban agriculture. What emerged in Cuba after conventional industrial agriculture became almost impossible in the 1990s were a number of new institutions and new policy measures that created a centrally directed organization providing motivation and administrative, technical, and material assistance to a decentralized network of small-scale agricultural production units in urban and semiurban areas.

In the process of making this shift in its agricultural sector, the government promoted activities in Cuban cities that are neatly summarized by two words: *urban* and *sustainable*. Concerning the first, there is little ambiguity in the concept of urban agriculture, which is primarily a territorial notion encompassing any agriculture practiced within or near population centers. A recent review article on urban agriculture in developing countries exemplifies such a purely territorial definition: "The words 'urban agriculture' will be used as defined by the growing of plants and the raising of animals for food and other uses within and around cities and towns" (De Bon et al. 2010: 21). Some authors have sharpened this definition by supplementing the "where" criterion with considerations of "by whom" and "for whom," stressing the reliance on "largely" local human and material resources and the "largely" local distribution of the products, services, and human and material resources generated (Mougeot 2000).

Regardless, Cuba, as we shall see, would have no difficulty in meeting the conditions of any conceivable definition of urban agriculture.

In addition, the definitions of urban and rural, and of what it means to be near versus far from a city, are essentially country-specific. In any country, these definitions are in general somewhat arbitrary, being driven by how the government delineates its territorial administrative units and by the existing transportation infrastructure and availability. Different answers to the question of how near is near would obviously introduce inconsistencies and complications into cross-country and even cross-study comparisons. These kinds of issues are of little concern to the task of this book, because Cuban authorities in charge of the urban agriculture effort have established painstakingly precise criteria in these matters (described in chapter 2), eliminating any ambiguity.

The concept of sustainability is somewhat more diffuse. It has a generic meaning of long-term persistence and reproducibility of any activity. In agriculture, the concept entered into scholarly and policy discussions in the late twentieth century as more and more people came to recognize that modern, industrial agriculture was giving every indication of being *unsustainable* in the long run. First, modern, industrial agriculture relies on extremely intensive use of fossil fuels, especially petroleum. Fossil fuels are required to run the machinery used in the large-field, mono-crop cultivation paradigmatic of industrial agriculture. Petroleum is also a key component of the chemical fertilizers and pesticides mandatory to produce the high yields demanded of this model of agriculture. Its products—being produced in massive quantities in relatively few locations and in excess of local needs—must be transported over long distances to reach consumers. This necessity introduces additional fuel use, not just in transportation but also in retarding spoilage of the foods during transport. M. King Hubbert's theory of a global peak in oil production with a subsequent downward trend once most global oil reserves have been located and the rate of discovery of new sources consequently declines has gained currency. This heightened awareness of the implications of profligate use of nonrenewable resources has raised questions about the long-term prospects for industrial agriculture. A second concern is the immediate, harmful effects on people and the environment of this way of producing foodstuffs. The agrochemical pesticides used on individual production units have widespread toxicity, not only to agricultural workers and consumers but also to animals and ecosystems. The runoff of excess fertilizer is polluting and degrading rivers, lakes, and bays. Soil conservation has become problematic due to the tilling techniques employed and the use of heavy machinery in the fields, which compacts the soil. And this catalog of drawbacks does not even mention the possible global environmental effects of greenhouse gas emissions and climate change.

In reality, it could be said that the discourse of sustainability in agriculture was born as opposition to "unsustainability" rather than as a coherent set of principles. Ideas about what it means to be agriculturally sustainable are quite varied. At the most basic level, a necessary and therefore relatively uncontroversial condition would seem to be maintaining (or increasing) yields while conserving the natural resource base of agriculture, all in a context of economic viability for the producers. Some have found this merely technical approach to the problem inadequate. In these opponents' view, "sustainable agriculture" makes sense only as a component of a more general notion of "sustainable development," a concept that incorporates considerations of inter- and intra-generational equity, of the provision of not only secure food supplies but all basic human needs for all people, of democratic participation in governance, and of the preservation and protection of the natural environment. As one author puts it, sustainable agriculture "cannot be usefully viewed merely or even primarily as farming systems that are technically able to maintain or increase yields while conserving their natural resource base. Such a narrow vision brushes aside crucial political, economic, and other social issues that have to be faced simultaneously with ecological ones at all levels from local to global" (Barraclough 2000: 9).

Agroecology emerged in the 1990s as a discipline that gives voice to these kinds of concerns. It applies ecological theory to agro-ecosystems, that is, plant-animal communities constructed and managed by humans to produce food, fuel, and fiber. These systems are studied in a holistic manner and managed in a way that pays attention to environmental concerns and to the human participants in their political, economic, social, and cultural settings. The agroecological performance criteria applied include ecological sustainability, food security (that is, individuals' access to adequate food supplies), economic viability, social equity and justice, conservation of resources, and increased production. Examples of such management techniques include eliminating the use of agrochemicals to reduce toxicity; increasing biodiversity and natural pest resistance by expanding the number of plant and animal species raised in one unit, and by practicing intercropping and crop rotation; and enhancing beneficial organic matter and biological activity in the soils being used for cultivation. The general goal is to manipulate agro-ecosystems to make them productive with fewer external inputs—ideally, in a self-sustaining manner—and with fewer damaging environmental and social outcomes (Altieri 1995, 2000).

In the early 1990s, history and force of circumstances induced Cuba to introduce urban agriculture as an important source of foodstuffs and as almost the only source of fresh produce in the diets of its city populations. Not only did families provision themselves through home gardens, but workplaces also initiated self-sufficiency production for their employees in nearby lots. Somewhat larger production units practicing innovative technologies began supplying their own workers as well as schools, hospitals, and a variety of different markets and produce stands selling food to urban residents. Circumstances also dictated that this turn to urban agriculture took place in a way that met the agroecological requirements for sustainability. In a way, Cubans were forced to act as trailblazers in unknown territory.

It is true that agriculture in cities has a long history and has been increasing in importance over recent decades throughout the world, especially in the developing world. But the urban agriculture movements taking place in Africa, Asia, and Latin America differ from the Cuban case in both their origins and their point of arrival. The florescence of urban agriculture in the third world in the second half of the twentieth century accompanied the massive population growth in its cities due to a hectic process of rural-urban migration. The peasants that migrated to cities in search of urban jobs and higher incomes ended up, more often than not, in the informal sector, living in poverty and with precarious employment or earnings. Either ignored or actively mistreated by urban governmental authorities, they saw growing food for themselves and for sale in urban food markets as a survival and income-generation strategy. The agricultural history, knowledge, and habits they brought with them from the countryside did not include any particular concern about use of fossil fuels, agrochemical pesticides, or fertilizers. They were not exactly steeped in agroecology. To this day they remain with one foot in the city and one foot in the countryside and rural habits. A recent comprehensive review of urban agricultural experiences in Asia, Africa, and Latin America reported that these urban agriculturalists continue to make use of petroleum-based fuels and agrochemicals (De Bon et al. 2010). Urban agriculture so implemented exacerbates hard-to-resolve problems of pollution and toxicity affecting the urban population. It also has to confront the reverse problem: pollution from other urban activities—industrial and human waste and emissions—degrading the urban soil, air, and water, and thereby degrading the quality of the produce itself.

In Cuba, by contrast, urban agriculture was not fueled by uncontrolled rural-urban migration. Nor, by the time the movement took place, were fossil fuels and agrochemicals readily available for use in agricultural production and distribution. Urban agriculture in Cuba was not simply a territorial rural-to-urban shift; it also involved a sea change in production technologies. It can and should be called a shift from industrial-rural to agroecological-urban agriculture using fossil fuels very sparingly as an energy source.

Another major contrast between urban agriculture in Cuba and in the rest of the world relates to the official, governmental recognition and support it received as it developed. In the developed world and in the postcolonial third-world countries that have inherited most of their ideas about appropriate urbanism from their former colonizers, agriculture in urban spaces is frowned upon. Thus, as rural migrants have descended on the cities and typically squatted on public and private lands to construct their houses and cultivate food crops, most governments have responded by ignoring or outright seeking to eliminate these developments. Neither governments nor private landowners approve of this helter-skelter land use, which conflicts with both official urban planning and private property rights. Zoning regulations, trespassing laws, and protection of property rights have rendered a considerable portion of urban agriculture illegal and subject to waxing and waning campaigns of eradication and forcible removal. This pattern of benign neglect alternating with periods of hostility persists in most places outside of parts of Asia and a few fledgling efforts elsewhere. China, with its long history of land scarcity, high population density, and production of food in or near cities, is perhaps the only exception to this generalization. In Cuba, by contrast, once urban agriculture was embraced as a potential way out of the food crisis, it enjoyed strong, continuous, and effective support from the Cuban government and leadership. One of the main arguments of this book, in fact, is that urban agriculture has been able to flourish in Cuba precisely because of the strong government support since the crisis, along with government policies in such areas as education, science, and technology over the 30 years preceding the crisis.

The same theme of benign neglect can be carried over to a discussion of the status of scholarly studies of urban agriculture. Mainstream academia has yet to produce book-length studies on the subject. Research and publication of books and case studies on urban agriculture have tended to take place in specialized research centers, development agencies, and nongovernmental and United Nations organizations focused on agriculture. A short list of such institutions would include the Canadian International Development Research Centre; the French Centre de Coopération Internationale en Recherche Agronomique pour le Développement (Center for International Cooperation in Agronomic Research for Development); the German Stiftung für Internationale Entwicklung (Foundation for International Development); the Oakland, California–based Food First (Institute for Food and Development Policy); and the international Resource Centres on Urban Agriculture and Food Security (RUAF). RUAF publishes *Urban Agriculture Magazine,* which by now contains an impressive list of case studies of urban agricultural experiences from all over the world.

Scholarly literature on Cuba's urban agriculture, especially in English, is more limited yet. José Alvarez (2004a) does an excellent job of covering the institutional and organizational measures and innovations Cuba instituted in the 1990s to restructure the production and distribution of agricultural output, but he addresses agriculture in general, not primarily the urban innovations. A number of articles relevant to but not directly focusing on urban agriculture have also appeared in *Cuba in Transition,* the annual published proceedings of the Association for the Study of the Cuban Economy (Espinosa Chepe 2006; Gayoso 2008; Hagelberg and Alvarez 2007, 2009; Royce 2004). However, there are no published book-length studies on Cuban urban agriculture as a whole, in either English or Spanish. One book-length study, *Agriculture in the City: A Key to Sustainability in Havana, Cuba* (Cruz and Sánchez Medina 2003), is very informative but addresses solely Havana. In fact, for the most part it focuses on urban agricultural projects in two small areas of the city: the Parque Metropolitano de la Habana, a large green area along the Almendares River that was designated a city park in 1963; and the 15.4 ha of urban agricultural activity in the municipality of Habana del Este (Cruz and Sánchez Medina 2003). In 1998 Food First also published an excellent monograph on agriculture in Havana titled *Cultivating Havana: Urban Agriculture and Food Security in the Years of Crisis*. It, once again, does not consider Cuba as a whole. Of course, by the time of this writing (2011) *Cultivating Havana* is becoming somewhat dated, as Cuban urban agriculture has continued developing at a rapid pace. One of the key leaders of the urban agricultural movement in Cuba, Dr. Nelso Companioni Concepción, has coauthored two short English-language pieces on urban agriculture in Cuba that cover the period up to the turn of the twenty-first century. These works, a 10-page journal article and a 17-page book chapter, deliver only a cursory treatment of the premises, organization, connection to sustainability, and results of urban agriculture in Cuba up to that time (Altieri et al. 1999; Companioni et al. 2002).

In summary, no existing published work has attempted a comprehensive, Cuba-wide consideration of the achievements of the Cuban urban agriculture movement with regard to production and all other aspects. There is a similar lack of analyses of the movement's results or the factors that made them possible. Cuban urban agriculture deserves more attention than that. The mere fact that, as we shall see, it managed to increase the production of vegetables in the cities a thousand-fold in the 12 years from 1994 to 2005 is prima facie evidence that it warrants a closer look.

To this end, chapter 1 outlines the historical circumstances of how urban agriculture got started in Cuba. Chapter 2 presents the institutional organization of the urban agriculture effort in Cuba. Chapters 3 through 6 discuss the enabling factors that contributed to the success of urban agriculture: research and development; training, education, and provision of inputs; material and moral incentives for the producers; and innovation and adaptation in technological implementation. Chapter 7 reports on field visits to particular urban agriculture production units, mostly in the city of Havana, Matanzas, and Pinar del Río. Chapter 8 turns to the consequences of urban agriculture, not only in terms of production but also in diverse areas such as urban employment, environmental restoration, community building, and gender equality. Chapter 9 places the Cuban experience in urban agriculture within the context of other experiences in Latin America and the Caribbean, and assesses the prospects for its future development.

I will argue that the factors underpinning the success of the Cuban urban agricultural effort can be broadly grouped into two categories:

1. *Organizational and policy innovation.* New organizational structures had to be established to connect the myriad of new, small-scale production units with each other and with central governmental authorities. New supply and distribution networks had to be set up to accommodate this new form of production. Technical training in new technologies and extension services had to be organized and successfully delivered to small, widely dispersed units of production. A variety of policies had to be adopted regarding, for example, material incentive schemes and growers' responsibilities to supply social service institutions such as schools, hospitals, and day-care centers. The result is a centrally guided system of production and distribution that functions as decentralized units.

2. *Human resources and capabilities.* Cuba had been building up its social capacity over the three decades separating the 1959 revolution and the onset of the 1990s crisis. It had made extraordinary progress at all levels of education, established research and development centers in a variety of scientific and technological fields, and sent graduate students to Eastern Europe and the Soviet Union to obtain their doctorates. Such efforts had created reserves of personnel highly skilled in sciences and technology. Cuba's successful universal education policy had produced a literate and numerate workforce capable of further education and retraining. All these conditions had positive implications for research and development, for technological innovation and adaptation, and for the staffing of relevant institutions.

After a thorough consideration of these enabling factors in the early chapters of this book, I turn to an assessment of the results Cuba has achieved in urban agriculture, including but not limited to gains in production of foodstuffs. The significant outcomes in increased employment, environmental recovery, and community building and strengthening are all worth noting and discussing. In fact, any serious evaluation of urban agriculture, wherever it is practiced, has to acknowledge its multifaceted nature and contributions (see, for example, Fleury and Ba 2005).

This assessment will be based primarily on information gathered during a three-month sabbatical stay in Cuba in the spring of 2007. During this stay I was able to interview prominent leaders in the urban agriculture community and visit urban agricultural production sites across Cuba, some of them multiple times, for direct observation and conversations with urban agriculturalists. I also interviewed Cuban scholars doing work on urban agriculture. During these interviews and visits I was given access to otherwise not easily available publications and data. These official sources were complemented by published data and information in secondary sources obtained during and after my stay in Cuba: Cuban publications, journal articles, newspapers, and information posted online by institutions of higher learning and research centers. I also obtained some data through direct observation, for instance, by recording food prices in various Havana food markets.

I would like to conclude this introduction with an observation on the potential global significance of the Cuban urban agricultural experiment. On the one hand, the Cuban case seems to be an extreme example. After all, the island was forced into urban agriculture and had no other viable alternatives. It simply had to stop using fossil fuels, machinery, and agrochemicals while increasing food production to meet the needs of its population. The limited quantities of agrochemicals, gasoline or diesel fuel, and machinery in working condition still available after the onset of the crisis precluded their use in the production of foodstuffs and in their transport to the cities—or even within the cities. As a result, Cuban urban agriculture does offer the world an example, probably the only extant one, of almost completely local and agroecological food production for local consumption. Its producer-to-consumer chain is probably the shortest in the world. Most of the produce is divided among the growers or sold to consumers on or next to the premises where it is produced. Urban agriculture has also become the main source of fresh produce for direct sale to the public; for provisioning social services such as schools, hospitals, correctional institutions, and workplaces; and for self-provisioning of families through home gardens. Cuban urban food is not just "organic," that is, produced without agrochemicals. It is also "local," obviating the need to use fossil fuels for transportation and for food preservation during transport (for example, via refrigeration or freezing). And, it is not a high-priced niche-market operation, as organic foods tend to be in the developed world. If and when environmental concerns and increasing resource scarcities push the rest of the world in the direction of agro-ecological food production and distribution, more and more people may find the pioneering efforts of Cuba in this area worth investigating, with a view to adapting components of successful Cuban practices to their own circumstances. Other third-world societies, the rapidly growing community-supported agriculture and "slow food" movements in the United States, and similarly motivated groups in other countries of the developed world will undoubtedly constitute receptive audiences for the lessons the Cuban experience has to offer.

1

◇◇◇◇◇◇◇◇◇◇◇◇◇◇◇◇

Cuban Agriculture

Historical Background and Key Concepts

As a member of the Council for Mutual Economic Assistance (COME-CON), the economic union of the Soviet Union and its allies, Cuba based its food production system on large, Soviet-style, industrial state farms. During this period, from the 1960s to about 1990, Cuban agriculture was moreover highly dependent on monoculture of sugar as an export crop to its COMECON trading partners, in exchange for a variety of generous Soviet economic support, as well as the petroleum and manufactured goods the island could not produce internally. With the rapid and dramatic collapse of the Soviet bloc in 1989–1991, Cuba almost overnight lost 85% of its export markets and its economic support, a challenge exacerbated by the U.S. economic blockade. Fuel to run agricultural machinery, spare parts and replacements for the machinery itself, and petroleum-based pesticides and fertilizers became essentially unavailable, precipitating Cuba into a national economic crisis officially dubbed the Periodo Especial en Tiempo de Paz (Special Period in Time of Peace).

Faced with increasingly serious food shortages, the Cuban government transformed the entire basis of its agricultural system, catapulting urban agricultural production from humble beginnings to robust performance in a remarkably short period. In order to understand how this transformation came about, it is helpful to consider the historical background and geographical setting that led to its development. Absent this history and geography—that is, if Cuba were not a tropical Caribbean island next to an imperial behemoth intent on dominating it—the crisis of the 1990s may not have occurred at all, or if it had, its course and Cuba's response to it would have been substantially different.

Two concepts of rather recent vintage are central to this discussion. The first is *food security*, referring to an individual's ability to access adequate, easily available food supplies on a regular basis (FAO 2006). The second, *food sovereignty*, insists that, to the extent possible, food security be based on locally grown foods whose production is controlled democratically by the farmers growing them, using local inputs and know-how, as well as ecological and sustainable agricultural practices (Desmarais 2003).

Cuban Food Security and Sovereignty during the Soviet Era

Neither history nor geography has been propitious for fulfilling either food security or sovereignty for the human populations of Cuba, or the entire Caribbean basin for that matter. The tropical climate makes it difficult to cultivate temperate-zone crops, such as wheat and soybeans, which are common staples of human and animal diets. Grains for feeding cattle or baking white bread, which is now central to the Cuban diet, must all be imported. In addition, the historical legacy of colonial agriculture—which was based on the cultivation of one or two highly labor-intensive export crops, mainly sugar, using slaves imported from Africa (and later indentured workers from China)—led to the relative neglect of food crops. Not only was land used disproportionately for export crops, but this relative overemphasis extended to areas such as research and development, credit and services provision, and governmental fiscal support. All these factors made food security import-dependent and likely to evaporate, especially for slave or slave-descendant populations, during hard times for export industries (Ahmed and Afroza 1996).

Still, by the end of the U.S.-dominated republican era—which lasted a little more than half a century, from the turn of the twentieth century to the victory of the Cuban Revolution over the forces of the dictator Fulgencio Batista in 1959—Cuba had achieved a respectable level of food availability by Caribbean and Central American standards, of 2,500 calories per capita per day. Food sovereignty was not, however, a realistic option, as the island remained very dependent on food imports (Nova González 2006: 173). Furthermore, it is highly unlikely that the food security situation was much better. Both land and income were very severely maldistributed. Poor rural Cubans, especially those employed only seasonally during the sugarcane harvest and left to fend for themselves in the *tiempo muerto* (dead time) following it, suffered greatly. The Cuban hinterland was significantly underdeveloped, and not only in terms of food production: 40% of the rural population was illiterate, less than 10% of rural households were electrified, and less than 3% had indoor plumbing. Only three general hospitals served all of rural Cuba prior to 1959 (MacEwan 1981: 19). These are not conditions conducive to widespread food security.

Beginning in 1959, the Castro government started implementing measures aimed at reducing the tremendous inequalities in income: agrarian and urban land reforms (the latter reducing rents), guaranteed employment, health care, and education. It also attempted to diversify agricultural production in the countryside away from heavy sugarcane cultivation. Unfortunately, this attempt proved technically ill-advised and had to be abandoned. The failure of this early diversification effort led to a renewed emphasis on sugar production, and especially on sugar-for-oil exchanges with the Soviet Union. This shift in orientation culminated in 1970 in a goal of producing 10 million tons of sugar. Although actual production fell short of that goal, the 8.5 million tons produced represented a historical record. Soon, sugar was again king in Cuban agriculture, but in a completely changed international context.

A significant feature of the government's reforms was nationalization of most privately held land in Cuba. The First Agrarian Reform, enacted by the new government in 1959, limited the maximum size of privately owned farmlands to 402 ha, and the Second Agrarian Reform (1963) further reduced the allowed holding to 63 ha. Any holdings above these limits were expropriated with compensation. These reforms naturally proved costly to U.S. businesses with property in Cuba, and relations between the two governments, already often strained, rapidly and irrevocably deteriorated. Cuba, for its part, declared itself socialist in April 1961 and sought protection in an alliance with the Soviet Union. By 1961 the United States had begun imposing trade restrictions on Cuba; these measures became increasingly severe, culminating in a total embargo by 1962. The United States also attempted unsuccessfully to invade Cuba in April 1961 in the notorious Bay of Pigs, landing a proxy army made up of Cuban exiles aided by the CIA. The two countries even approached the brink of nuclear disaster during the Cuban Missile Crisis of 1962.

By 1962 shortages began developing in the markets, so the Cuban government introduced food rationing in order to distribute supplies equitably in a way that enhanced food security for the entire population in a context of shared hardships (MacEwan 1981: 65). Rationing and a ration book (*la libreta*) that entitles every Cuban to a basket of food and household items at a nominal cost have been permanent features of the Cuban economic landscape ever since, although in the Special Period ration entitlements have dropped dramatically.

Alignment with the Soviet Union was to have fateful consequences for Cuba. In the 30 years spanning 1960 to 1990, Cuba became closely entwined with the Soviet bloc, not only politically and militarily, but also economically. In 1972 Cuba joined COMECON and started fulfilling the obligations assigned to it within the framework of this organization. In the agricultural area, Cuba essentially became the provider of sugar and citrus fruits to COMECON. In return, it received those items needed to sustain and expand a national economy for which it lacked sufficient production capacity, primarily manufactured (including capital) goods, petroleum, and processed food. The terms of trade received by Cuba were quite favorable, entailing a substantial COMECON subsidy for the country.

The close affiliation with the Soviet Union had other consequential effects on the Cuban agricultural sector, which took on most of the characteristics of the centrally planned Soviet agricultural system. By 1975 nearly 80% of agricultural lands in Cuba had been incorporated into the state sector and organized into gigantic farms under centralized government control. In 1987 there were only 474 state farming enterprises in all of Cuba. The average size of these state farms was 17,400 ha, and they focused on sugarcane or rice cultivation, cattle ranching, and forestry enterprises. Even the comparatively small *cultivos varios* units growing potatoes, grains, vegetables, and tubers had a mean size of 4,600 ha (Valdés Paz 1997: 149).

All these enterprises practiced industrial agriculture, which relies heavily on machinery, chemical pesticides and fertilizers, and transportation networks. Both the petroleum (as well as petroleum-based agricultural chemicals) and the machinery (and necessary spare parts) had to be imported. By the end of the 1980s, Cuba was using more fertilizers per hectare than the United States or any Latin American country: 202 kg/ha compared with 93 kg/ha in the United States and 56 kg/ha in Latin America. Its use of 22 tractors per 1,000 ha exceeded the averages for the Caribbean region (17), Latin America (11), and the entire world (19) (Nova González 2008g; FAO 2003).

Over the 30 years following 1960, Cuba did achieve food security for its population using this model of agricultural production. By 1989, per capita calorie availability had reached nearly 2,900 calories per day, a figure well above not only the current daily minimum requirements of the Food and Agriculture Organization (FAO) of the United Nations (about 1,900 calories per day) but also of the FAO daily recommended quantities (about 2,400 calories per day) (Nova González 2006: 180; FAO 2009). This high average nutritional intake was coupled with full employment in the economy, entitlements to food through the libreta, and free meals provided in schools and workplaces, factors that point to substantial food security for the Cuban population during this postrevolutionary period. The food sovereignty situation was substantially different, however. Cuba remained extremely dependent on imports and thus vulnerable to interruptions and shortages in the food supply. In 1986 Cuba imported—mostly from COMECON—54% of the calories and 61% of the protein its population consumed (Nova González 2006: 181). Compounding the problem, local production of foodstuffs on the island was almost totally dependent on imports of petroleum, petroleum derivatives, and agricultural machinery.

The Economic Crisis of the Special Period

Fragility and vulnerability in the food supply were inherent in this combination of more-than-adequate food security and heavy reliance on imports, a danger that began to manifest in the late 1980s. By that time Soviet-style inefficiencies had led to increasingly lethargic productivity gains in agriculture and throughout the Cuban economy. The real crunch, however, came in 1989–1991, a three-year period when the Soviet Union ceased to exist, and along with it, the Soviet-led military (Warsaw Pact) and economic (COMECON) alliances in Eastern Europe. Cuba, almost the only country in the former Soviet bloc that did not voluntarily abandon its socialist model and principles, was left facing an acute economic crisis that threatened the very survival of its population. It was, figuratively speaking, as if a dam had burst. Cuba lost all of its external trade and economic-aid relations, relations that had been carefully cultivated by the Cubans and the European socialist states, especially the Soviet Union, ever since the beginning of the Cuban Revolution and that had become dominant in Cuban life in the 1970s and 1980s, following full Cuban membership in COMECON. Their precipitous removal caused an immediate crisis in both the agricultural and nonagricultural sectors of the Cuban economy. As a result the Cuban government declared in 1990 that Cuba had entered a Special Period in Time of Peace, requiring its citizens to make the kinds of sacrifices usually imposed in wartime.

Simultaneously with this crisis, and continuing through the 1990s, the island came under increasing pressure from hostile U.S. economic policies designed to paralyze the Cuban economy in the hopes Castro would be overthrown. Key among these were the Torricelli Act (1992), which, among other provisions, banned exports of food and medicine to Cuba (excepting only humanitarian aid), and the Helms-Burton Act (1996), which subjected foreign corporations doing business in Cuba to U.S. sanctions (Espinosa Martínez 1997). These measures further strangled Cuban economic output. Estimates of the decline in the Cuban gross domestic product (GDP) between 1989 and 1993 range from 35% (García 1996: 44) to 50% (Espinosa Martínez 1997: 9). During this time, petroleum imports dropped from 13.3 to 5.37 million metric tons (t) per year, and exports and imports in general suffered declines of more than 80% (Espinosa Martínez 1997: 10; Koont 1994: 6). Most severely affected was the manufacturing of non-food consumer goods, as factories ceased production for lack of energy and raw material inputs. Because the state was no longer able to offer people consumer goods that could be purchased in pesos, several related problems developed in the macro-economy. First, the Cuban pesos in which workers were paid became essentially unspendable, undermining labor discipline and efficiency as earning pesos became rather irrelevant. Second, black and gray markets emerged, trading in goods pilfered from the state, in which the currency of exchange was U.S. dollars, either earned from the tourism sector or sent to Cuba as remittances from relatives exiled in the United States. This informal economy created glaring socioeconomic disparities, as those Cubans with access to U.S. dollars could obtain products unavailable to the rest. And third, the Cuban peso, officially on par with the U.S. dollar but not freely convertible, depreciated to a level as low as 150 pesos to the dollar in the black market.

Agriculture virtually collapsed as the Cuban economy imploded. Available data underline the gravity of the situation: from 1989 to 1995, fertilizer use decreased from 900,000 to 250,000 metric tons. In 1989 the agricultural sector had 736 million pesos' worth of basic inputs available,

Table 1.1. Per capita Cuban agricultural production averages

Commodity	Average per capita production (kg/year)	
	1986–1990	1991–1995
Roots and tubers	95	89
Vegetables	55	38
Cereals and legumes	54	31
Fruits	108	73
Cow's milk	99	46
Total meat in distribution	41	24
Eggs (in numbers)	241	164

Source: Nova González 2006: 72.

but by 1993 that amount had plummeted to 204 million pesos (Nova González 2006: 72). Table 1.1 summarizes what happened to per capita production levels under these circumstances.

By 1994 the daily per capita nutritional intake of the Cuban population had reached its nadir at levels well below the FAO recommendations for a healthy diet (given in parentheses): 1,853 calories per day (2,400 calories recommended), 46 grams of protein (72 grams recommended), and 26 grams of fat (75 grams recommended) (Alvarez 2004b: 3; Nova González 2006: 198).

Of necessity, the Cuban state adopted a series of unprecedented measures to confront the catastrophic effects of the collapse of COMECON and the increasingly hostile U.S. economic policies. In particular, three major steps were taken in 1993. The first two were, in effect, emergency measures that legalized and regularized certain previously banned economic activities that were already taking place in the burgeoning Cuban black and gray markets: first, the possession and circulation of U.S. dollars by Cuban citizens was legalized; and second, self-employment in a wide list of professions was made legal and subjected to licensing and taxation. Both forms of economic activity promoted remittances of U.S. dollars from abroad and filled gaps in the provision of goods and services to the population created by the state's faltering production capacity. Regularizing their status and making them subject to taxation in various ways therefore made eminently good sense. In any case, both were emergency measures forced upon the state by circumstance and subject to future reversal as conditions allowed. In fact, Cuba revoked the dollar's legal status as a circulating currency in 2006. A dual monetary system lives on, however, in the continuing presence of a "strong currency," the so-called *peso convertible* (CUC), which is more or less equivalent to the U.S. dollar, in tandem with the ordinary Cuban peso, the so-called *moneda nacional,* with the state maintaining a sort of a controlled convertibility between the two.

These first two measures were not directed at agriculture, although they certainly affected it. In contrast, Decreed Law 142, adopted by the Council of Ministers on September 20, 1993, specifically addressed the agricultural sector, introducing substantial and difficult-to-reverse institutional changes. These were so significant that analysts have dubbed the measure the Third Agrarian Reform, on a par with the first two structural and transformational land reforms shortly after the revolution. The very title of this decree introduced a new institution and a new name into Cuban agriculture: Decreto-Ley Número 142 Sobre las Unidades Básicas de Producción Cooperativa (Decreed Law Number 142 on the Basic Units of Cooperative Production).

This law essentially acknowledged that, as a result of COMECON's collapse, a system of production based on state farms had simply become impossible. Its first article announces the breakup of state farms under the control of the Ministerio de la Agricultura (MINAG, Ministry of Agriculture) and the Ministerio de Azúcar (MINAZ, Ministry of Sugar), and their replacement by new, much smaller cooperative productive units, called Unidades Básicas de Producción Cooperativa (UBPCs). State workers employed in the gamut of agricultural activity—animal raising as well as cultivation of sugarcane and non-sugar crops—were simply to be converted into cooperative members.

The idea underlying the law is simple: to stimulate production by encouraging local autonomy in decision making and by linking the incomes of cooperative members not to a centrally determined salary scale but to actual production levels achieved by the cooperative. Article 1 of Decreed Law 142 lists the underlying principles sustaining UBPC activities:

- Each worker is to be attached to a specific piece of land.
- The workers in the collective are to provide themselves and their families with their own food, improving their quality of life.
- Workers' incomes are to be tightly tied to the achievement of specified production levels.

- The development of a high degree of management autonomy is to be encouraged. Each UBPC is to administer its own resources, becoming self-sufficient in production.

Article 2 specifies that the UBPCs will be given the following legal rights, vested in them in the same terms as in an individual person:

- The UBPC has usufruct rights to a plot of land for an indefinite time.
- The cooperative owns all that is produced on that plot.
- Harvests beyond the amount needed for self-provisioning of workers can be sold to the state, either directly through the UBPC enterprise or in a manner decided by the cooperative.
- The UBPC pays for its own technical and material supplies.
- For this purpose it has a bank account.
- The UBPC can buy on credit the fundamental inputs required for production.
- Collective members elect their own leaders, who must give periodic accountings to the membership.
- The UBPC must comply with its fiscal obligations to contribute to the general expenditures of the nation.

In addition to these two articles, the law contains a "special disposition" for small, isolated plots of land. These plots, limited to 0.5 ha in size, are to be made available in indefinite usufruct to retired persons, or to other persons unable to work in agriculture on a regular basis for justifiable reasons, so that they can grow food for themselves and their families. The same disposition allows isolated plots formerly used for growing tobacco but now abandoned to be given in usufruct to individuals willing to work them, whether or not they are retired (Valdés Paz 1997: 247).

The implementation of Decreed Law 142 was swift. By the mid-1990s more than 3,000 UBPCs were functioning, about half growing sugarcane and the rest focusing on other areas of agriculture. Yet the breakup of state farms and the alteration of economic incentive structures did not by themselves immediately translate into successful harvests. Even after the organizational restructuring of the production process, the necessary inputs—agricultural machinery and spare parts, fuel to operate machinery and for the transportation of goods, and petroleum-based fertilizers and pesticides—continued to be in critically short supply. Only a fairly small minority of the newly organized UBPCs were able to achieve profitability based on efficient production and high productivity (Nova González 2006: 164). In sum, the Third Agrarian Reform significantly altered the organization of the production process and the nature of land tenancy (i.e., the conditions of "property ownership") and seemingly became a permanent part of the Cuban economy, since its provisions would be extremely difficult to reverse. It did not, however, resolve all the difficulties facing Cuban agriculture (Alvarez 2004a: 75–87; Royce 2004).

The Turn to an Agroecological Approach

Cuba's population is overwhelmingly urban, with more than 90% living in and around cities. The reforms of 1993 left essentially unresolved how to supply this majority with adequate nutrition, given the chronic shortage of petroleum for food transportation. A way of growing and distributing food that minimized use of petroleum-based inputs was urgently needed. In this context, it is not surprising that urban agriculture presented a possible solution. Growing food near where people live eliminates the need to use gasoline and diesel fuel for transportation. Shortening the time lag between harvest (or slaughter of livestock) and kitchen also reduces spoilage and quality deterioration in the foodstuffs.

What tipped the balance from possibility to feasibility of urban agriculture was the unavailability of the chemical pesticides and fertilizers used in industrial agriculture. The large-scale use of such toxic substances in close proximity to densely populated areas is impractical and dangerous to public health. Thus, urban agriculture turns to organic or agroecological alternatives, such as the use of organic manures, compost, and worm humus (nutrient-rich worm excreta); biological pest control; and little use of machinery. In fact, it makes little sense to distinguish between "urban" and "organic" agriculture in Cuba, as almost all organic agriculture is practiced in urban areas, and almost all urban agriculture is organic. Cuban farmers, not to mention the Ministry of Agriculture, would doubtless have resisted the island-wide forced adoption of such techniques on a large scale had petroleum-based options been available. Yet the absence of such options dovetailed very well with the prospect of introducing large-scale urban cultivation of food crops. Moreover, the higher labor intensity associated with ecologically friendly techniques had a better chance of being accommodated in cities than in the countryside. Highly educated urban populations left effectively unemployed by the shrinkage in the economy formed a reservoir from which large numbers of urban farmworkers could be recruited.

Yet during the 1980s, even before the crisis of the Special Period, glimmerings of an agroecological mind-set, research, and practice had appeared in Cuban agriculture. The first harbinger was growing dissatisfaction with the performance of enormous Soviet-style state farms practicing industrial agriculture. This dissatisfaction was one of the earliest expressions of the current worldwide awakening and evolution toward the advocacy and practice of ecological, sustainable agriculture. In Cuba's case, by the late 1980s the centrally planned, *latifundio*-style monoculture farming of sugar and other crops was facing increasing physical and organizational limitations, leading to stagnation, if not in absolute production at least in productivity growth.

On the labor front, rural populations continued declining both in absolute terms and relative to urban populations, making it increasingly difficult to attract a stable workforce (Valdés Paz 1997). The net output per worker diminished steadily from 2,286 pesos per worker in 1981 to 1,770 pesos per worker in 1989 (Nova González 2008g). One contributing factor was doubtless the gradual replacement by salaried workers of the traditional peasant farmers, who felt connected to the land they were working and possessed a store of accumulated knowledge about farming. Ecological issues compounded the labor problem. The technologies inherent in large-scale industrial cultivation had led to the consolidation of farming lands into large, contiguous units devoid of trees. The use of heavy agricultural machinery—mainly tractors—and the liberal application of imported fertilizers, pesticides, and herbicides severely degraded the cultivated soil. By the early 1990s, excessively deep plowing; heavy use of herbicides; and lack of appropriate crop rotation, intercalation (i.e., mixture of crops in close proximity, intercropping), and diversity had led to erosion on 65% of agricultural land, with 25% of the land experiencing severe or very severe erosion. In addition, heavy machinery had compacted the soil on 2.5 million ha of land. Related problems with poor drainage (affecting 1.5 million ha), soil salinity (affecting 1.0 million ha), and acidity (affecting 1.1 million ha) also became evident. Only 25.2% of Cuban agricultural lands could be classified as very productive or productive (Nova González 2008g).

Of course, this accumulation of negative factors could not help but be reflected in efficiency indicators for the agricultural sector. Between 1976–80 and 1986–90, the output-to-capital ratios in the sector diminished from 5.5 to 3.9, the investment required to produce 1 peso's worth of product rose from 0.18 to 0.26 peso, and the material inputs needed for 1 peso's worth of gross production rose from 0.56 to 0.68 peso. This does not mean that agriculture was starved of investment. In fact, it received about 30% of all investment in the Cuban economy in the decade from 1980 to 1989, more or less equally divided between the sugar and non-sugar sectors (Nova González 2008g). All these problems led to a growing realization that Cuba's chosen path had not only increased its dependence on imported inputs and caused severe environmental degradation, but also led to smaller harvests per peso of investment.

Some voices began to argue for more ecologically friendly growing techniques that used locally produced biological inputs in place of chemicals and relied on farmers' traditional knowledge and practices in crop rotation, intercropping, and crop diversification in settings reasonably scaled for labor-intensive cultivation. One of the first official and public responses to these arguments was the registration in February 1987 of the Asociación Cubana de Técnicos Agrícolas y Forestales (ACTAF, Cuban Association of Agricultural and Forestry Technicians) as a nongovernmental organization affiliated with the Ministry of Agriculture. The aim of ACTAF is to support the development of sustainable agroecological agriculture (ACTAF 2004), and as we will see, it became a major player in the development of urban agriculture. ACTAF recruits dues-paying members from among agricultural professionals and technicians, retirees with an interest in its work, and small farmers—the last with the permission of the Asociación Nacional de Agricultores Pequeños (ANAP, National Association of Small Farmers). Institutions and organizations may also choose to be affiliated. Although ACTAF did not become highly active until the onset of the Special Period crisis, the fact that it was organized in 1987 indicates that at least some Cuban agronomists were already beginning to rethink industrial agriculture.

The second concern leading to an agroecological mind-set in Cuba was rather more specific to Cuba, namely, Cuba's relations with the United States. Starting in the early 1800s (with Presidents Jefferson and Monroe) and continuing to this day, the government of the United States has done everything possible to deny Cuba the status of a truly independent country (Morales 2008). During the nineteenth century, when Cuba was still a Spanish colony, the United States worked against Cuban independence from Spain, even as the rest of Latin America successfully founded independent nations in the first third of the nineteenth century. The United States then tried to purchase and annex Cuba, and its southern filibusters staged armed invasions to bring about Cuba's entry into the nation as a slave state. At the end of the Spanish-American War, Cuba, Puerto Rico, and the Philippines found themselves under U.S. control. The Cuban Republic instituted at that time was in fact more like a quasi-colony of the United States. Following the Cuban Revolution in 1959, Fidel Castro's actions to assert true independence and to insist on Cuban sovereignty provoked a vehement U.S. response. The Eisenhower administration tried to prevent the new revolutionary government from taking power and, once it did come to power, immediately began working to overthrow it.

Since that time, the United States has maintained a consistently hostile posture toward Cuba, including overt military action (the Bay of Pigs), attempts at internal sabotage (Operation Mongoose), and the previously mentioned economic sanctions. The U.S. government has clearly stated the unacceptability of the Castro government, and has maintained unrelenting propaganda broadcasts via Radio Martí and TV Martí. Other discriminatory and hostile legislation in effect in the United States restricts rights of migration and travel between the two countries. In this climate, the Cuban government would have been naive and extremely foolish if it did not consider the possibility of a complete U.S. blockade of the island that would cut off all imports from the Soviet Union. Shortly after the Castro revolution, Cuban scientific institutions started researching possibilities for import substitution in production, including agricultural production, in order to make the island less dependent on imported goods. At the same time, the Ministerio de las Fuerzas Armadas Revolucionarías (MINFAR, Ministry of Revolutionary Armed Forces) and institutions such as the Instituto Nacional de Reserva Estatal (INRE, National Institute of State Reserves) began studying potential responses to a complete cutoff of petroleum imports. (Interestingly, the Ministry of Agriculture, which was very committed to industrial, high-input agriculture, was not part of this effort.)

Raúl Castro is credited with providing the first official government encouragement of low-input agricultural production. During a visit to the Armed Forces Horticultural Enterprise (HORTIFAR) on December 27, 1987, then Minister of Defense Castro told of his encounter with an agricultural engineer named Ana Luisa Pérez, who had carried out some successful experiments growing vegetables without using petrochemicals in plantain pregerminators.[1] Castro suggested the desirability of generalizing this method of cultivation, a comment that led to the introduction of a technology later widely employed in urban agriculture (Castro Ruz 1997). In response, Armed Forces facilities began installing so-called *organopónicos*—raised cultivation beds containing a mixture of soil and organic material such as compost—beginning in December 1987, four years before the demise of the Soviet Union. The first organopónico, with an area of 1 ha, was constructed next to the base of the Fiftieth Division of the Armed Forces in eastern Cuba, built by family members of the officers and soldiers of the division. Apparently, one impetus was that the commanding general of the division, General Néstor López, was an old campesino who enjoyed growing food (General Moisés Sio Wong tells this story in Choy et al. 2005: 129).

By the end of 1991, the first organopónico not located in a military unit was constructed in a vacant lot next to INRE's headquarters in the Miramar district of Havana (fig. 1.1). The impetus for this installation came from the Fourth Party Congress of the Cuban Communist Party, held in Santiago de Cuba in October 1991. It was at this congress that the label "Special Period in Time of Peace" was first applied to underline the gravity of the Cuban economic crisis. Food production was singled out as the most pressing need, leading to the adoption of the Food Program (Programa Alimentario), an effort focusing on the immediate production of foodstuffs in every possible location, using available local resources. In Havana, General Sio Wong, a descendant of Chinese immigrants brought to Cuba as indentured workers in the sugarcane fields after the abolition of slavery, took up the challenge. In an interview published in *Agricultura Orgánica,* he gave three reasons for becoming personally involved in the Food Program. He was convinced that vegetable production in the cities, especially in Havana, would prove the quickest and most effective response to the food emergency. He also believed that urban agriculture could play an important role in providing domestic supplies for the tourism sector, slated for rapid growth in the coming years. The third reason was wrapped up in his history and ethnicity. It had been the Chinese indentured workers who had introduced vegetable cultivation in the

Figure 1.1. INRE-1 in the city of Havana, the first organopónico installed outside a military base. (Photo by author.)

communities where they lived. The general Cuban population lived on a diet based on pork, rice, and beans, in which vegetables played little part. According to Sio Wong, the Chinese possessed the culture and know-how to become successful vegetable growers, such that in most towns "one Chinese could produce enough vegetables for the whole town" (Carrión Fernández 2006). Undoubtedly as well, his position as head of INRE, the emergency preparedness agency of the state, made it nearly inevitable that he would get involved.

Shortly after the Fourth Party Congress, an assembly of the workforce at INRE agreed to construct an organopónico in an adjoining empty lot (Choy et al. 2005: 129). In December 1991 Sio Wong sought technical help in launching INRE's organopónico. He paid a visit to the Instituto de Investigaciones Fundamentales en Agricultura Tropical "Alejandro de Humboldt" (INIFAT, Alexander Humboldt Institute of Fundamental Research in Tropical Agriculture), located at Santiago de las Vegas, 25 km south of Havana in the province of La Habana. Formed in 1974 as an amalgamation of three institutions within the Academia de Ciencias Cubanas (ACC, Cuban Academy of Sciences), INIFAT was a well-established research center.[2]

Thus, with the impetus from the Fourth Party Congress and the technical help of INIFAT, Sio Wong and his team began transforming their vacant lot into a vegetable plot. Political support from Raúl Castro was important to their efforts, because they encountered significant resistance from local authorities, who saw the use of concrete in the construction of the walled beds as ridiculously wasteful in the context of extreme shortages in building materials. The INRE workers proceeded to harvest the first crop of lettuce grown in an organopónico on January 28, 1992 (Carrión Fernández 2006). Since then, the organopónico has become one of the mainstays of vegetable cultivation in Cuban urban agriculture.

When the decision was made to extend the practice of urban agriculture, including the new technology of organopónicos, to all urban areas in Cuba, steps were taken to give the effort an organizational backbone at the highest levels of government. By the end of 1993, the provincial government of the city of Havana and the Ministry of Agriculture both were already strongly supporting extension efforts to spread the new technology throughout the capital city. The first organizational push at the national level came in May 1994 at the First National Plenary of Organopónicos in Santa Clara. At this meeting a national commission was selected to organize and guide the organopónico movement across the nation. This national commission, named the Movimiento Nacional de Agricultura Urbana (National Movement of Urban Agriculture), was to be headquartered at INIFAT with that institute's two top leaders, Dr. Adolfo Rodríguez Nodals and Dr. Nelso Companioni Concepción, serving as president and executive secretary, respectively. General Sio Wong was a member, as were a number of other prominent Cuban agriculturalists, including the then head of the Provincial Agriculture Group of the City of Havana (Grupo Provincial de Agricultura en Ciudad de la Habana), Eugenio Fuster Chepe, who would later become the president of ACTAF, the aforementioned nongovernmental organization dedicated to sustainable agriculture.

The final form into which this organizational structure evolved is discussed in the following chapter. But what is clear is this: by the time the Special Period crisis made the shift of agricultural production to cities unavoidable in the early 1990s, at least some components of the Cuban institutional structure were able to harness technologies, policies, and practices that had already been in gestation for a number of years. When forced to move in the direction of self-sufficiency and agroecological technology, Cuba was not only prepared, but it started formulating immediate, large-scale, and effective responses within its national governance structures.

2

◇◇◇◇◇◇◇◇◇◇◇◇◇◇◇◇

The Nature and Organization of Cuban Urban Agriculture

Urban agriculture, as currently defined and practiced in Cuba, is essentially nothing more than small-scale agriculture practiced near urban populations and using very little petroleum (or petroleum derivatives) and machinery. The Grupo Nacional de Agricultura Urbana (GNAU, National Group for Urban Agriculture), the top overseers of the urban agriculture movement, describe it as follows: "the production of food within the urban and peri-urban perimeter, using intensive methods, paying attention to the human-crop-animal-environment interrelationships, and taking advantage of the urban infrastructure with its stable labor force. This results in diversified production of crops and animals throughout the year, based on sustainable practices which allow the recycling of waste materials" (GNAU 2007a: 4; my translation). Dr. Nelso Companioni Concepción, secretary of the GNAU and one of the organizers of the urban agriculture movement, describes urban agriculture as a popular movement of food production with a strong base in sustainability, in which the producer, as the principal actor in the entire production process, has total responsibility for its management, including the marketing of the product (Companioni Concepción 2007, slide 2).

The "urban" designation in the name is in fact a bit of a stretch, as some pretty rural-looking suburban and peri-urban expanses are defined as urban agricultural lands. The GNAU's formal definition of urban locations includes all agricultural lands within certain distances of cities and towns having populations in excess of 1,000 persons (see table 2.1). The entire province of Ciudad de la Habana is also considered urban, as are any gardens families use for growing their own food, provided they live in a settlement of at least 15 houses. Nearly 9 million Cubans currently

Table 2.1. Extent of urban agricultural lands in Cuba

Territory	Geographical extent of urban agriculture	Total area (ha)
La Habana	Entire province	35,902
Provincial capitals and Manzanillo	Within 10 km of the city	163,363
Municipal capitals	Within 5 km of the city	637,834
Other cities and towns	Within 2 km of the city	380,566
Settlements	Self-provisioning family gardens	45,578
Total		**1,263,243**

Source: Companioni Concepción 2007: slide 6.

live in the 761 cities and towns that have populations of more than 1,000. Another 2.25 million reside in the 6,427 settlements with between 50 and 1,000 inhabitants, a substantial proportion of which would be included in the urban agriculture statistics. In all, about 95% of the Cuban population lives in environments defined as "urban" under these criteria. It should not be surprising, then, that the total urban agricultural land area is far from negligible, estimated at more than 1.25 million ha and accounting for 14.6% of all agricultural lands in Cuba (Companioni Concepción 2007, slides 5–6).

The second primary feature of urban agriculture, its sustainability, goes to the very nature of the kind of agriculture that can be practiced in or near population centers. Because of the deleterious effects of chemical fertilizers and pesticides on human health, agricultural activity in or near cities must minimize their use, both directly and indirectly through contamination of soil and water. It is important to bear in mind, however, that the turn toward agroecological, sustainable, and environmentally friendly technologies was initially forced on Cuba by its inability to import petroleum and petroleum derivatives following the collapse of COMECON. Such imports as could still be afforded were needed for crops such as potatoes and sugarcane, for which adequate alternative technologies were not available. It was in these circumstances that Cuba decided to look for a silver lining in a very dark economic cloud and initiate an urban agriculture that, in its very conception and guiding principles, sought to be an example of urban sustainability. Large-scale industrial agriculture, although deeply entrenched in the structure and mind-set of the Ministry of Agriculture, had simply become impracticable and had to be replaced by a small-scale, labor-intensive form of agriculture based on quite different technologies and types of social organization.

Companioni emphasizes that the organic character of urban agriculture did not grow out of a desire to market the produce as organic, a frequent motivation in the developed world, where organic products command much higher prices in niche markets. Rather, it followed from the unavoidable necessity of using only locally available (or locally developable) resources in the production process. This constraint has imparted on Cuban urban agriculture its endogenous and locally sustainable nature. Local sources must be found for the provision of organic fertilizers, biological pest-control substances, seeds, tools, training, other services, and even the labor force.[1] The strategic framework for this movement is encapsulated in the following five key features (Companioni Concepción 2007, slide 13):

- The use of mechanisms that create the conditions for successful production and give adequate incentives for increasing production, including adequate remuneration, extension services, organic materials, seed and seedlings, irrigation, and biological pest control.
- The deliberate and intensive use of all agricultural land available in and near cities and settlements, with a defined program for each production unit, and systematic supervision and inspection of all activity.
- Maximum diversification of plant species, animal breeds, and varieties in each production unit.
- A raised level of public consciousness and culture concerning agriculture, nutrition, and the environment through a dynamic extension service that provides all production units with ongoing technical training, disseminates information about new scientific and technical advances, and publicizes the positive experiences of successful producers.
- A close cooperation among all entities (scientific, educational, productive, and social-services oriented) involved in the production, processing, and distribution of foodstuffs, including the Communist Party of Cuba, the government, and mass organizations (such as trade and student unions and the women's federa-

tion); the food producers are in the central position as the direct
and principal actors in distributing and selling their own harvests.

According to Eugenio Fuster Chepe (2006), president of the nongovernmental organization ACTAF and a prominent leader in this movement,
the development of urban agriculture in Cuba is based on six fundamental concepts: agroecology and sustainability, diversification of production,
small-scale cultivation, adequate economic incentives to producers, harmony with the urban surroundings, and the war of the whole people. The
last point refers to the strategic conception for national defense elaborated
by MINFAR. It is based on the most varied and efficient use of all material
and moral resources of the Cuban society in confronting enemies that
are numerically and technologically superior. It once again brings into
focus the national security and defense concerns driving this agricultural
transformation. The Cuban government believes it to be vital that in a
crisis, such as aggression against and possible invasion of the island, Cubans must be able to feed themselves with locally produced foods. Fuster
Chepe (2006) identifies the four key premises that underlie the Cuban
urban agricultural effort: first, the bringing together and organization of
all who practice agriculture in and near cities; second, production in the
neighborhood, by the neighborhood, and for the neighborhood; third,
provision of preparation, training, and assistance to food producers, in
order to ensure their success; and, fourth, most importantly, decentralization of all links in the chain from planting to plate: the supplying of inputs,
production, and marketing.

As this discussion makes clear, urban agriculture in Cuba is more than
just a matter of growing food in and near cities. It embodies a whole new
approach to mass production of foodstuffs that minimizes harmful impacts on the environment and ensures safe, wholesome, and healthy conditions for all urban residents, whether they are agricultural workers or
food consumers. Importantly, a great deal of local autonomy is invested
in the growers themselves, because they decide what to produce and how
to distribute it.

Another significant change driven by the urban agriculture movement
is in the nature of landholding. The First and Second Agrarian Reforms,
in 1959 and 1963, respectively, nationalized large landholdings and created
two basic types of agricultural workers. The first type was small landowners working privately owned land held either in a farmers' cooperative or

Table 2.2. Agricultural land tenancy in Cuba, 2008

Name of unit	Form of tenancy	Status of tenancy	Description
Cooperativa de Producción Agropecuaria (CPA)	Collective	Private	A cooperative formed by farmers who merge their individual holdings into the cooperative
Cooperativa de Créditos y Servicios (CCS, CCS Fortalecida)[a]	Mixed collective and individual	Mixed private and usufruct	Land held by individual small farmers who have joined a cooperative to obtain credit and services
Unidad Básica de Producción Cooperativa (UBPC)[b]	Collective	Usufruct	A cooperative of individuals who join together to collectively farm lands made available to them by the state in usufruct
Parcela[b]	Individual	Usufruct	Plot of land obtained by an individual in usufruct from the state for the purpose of growing food
Patio[b]	Individual	Private	Home garden planted for personal consumption or small-scale sale
New state farms (fincas and empresas)[c]	State	State	Farmlands that have stayed in the hands of the state and been reorganized into smaller units growing crops for public consumption and export
Autoconsumo units for state enterprises	State	State	Self-provisioning gardens for the workers of a particular state enterprise
Ministerio del Interior (MININT)	State	State	Interior Ministry–managed lands used in meeting the ministry's self-provisioning needs
Ministerio de las Fuerzas Armadas Revolucionarías (MINFAR)	State	State	Land controlled by the Armed Forces used for provisioning military units with food

Notes: a. CSS Fortalecidas have more formal structure and organization (such as strict accounting practices) than regular CSSs. Any CSS can become fortalecida by meeting set requirements.
b. New landholding forms introduced by Special Period reforms. Home gardens (*patios*) have always existed, but the government now strongly promotes and supports them.
c. The enormous farms of pre-1990 have been broken into smaller units and undergone restructuring, hence the "new" designation.

individually. These are Cooperativas de Producción Agropecuaria (CPAs, Agricultural Production Cooperatives) and Cooperativas de Créditos y Servicios (CCS, Cooperatives for Credits and Services), respectively. In a CPA, the members combine their lands into joint cooperative ownership, whereas in a CCS, land continues to be held individually either in outright ownership or in usufruct. The cooperative allows the farmers jointly to obtain credit and other services.[2] The second type of workers were state employees working on public lands in various state enterprises (see table 2.2). These landholdings continue to exist, although with sizes and structures modified by Special Period laws. The tenancy structure has, however, become more complicated since the early 1990s, with the addition of UBPC members and individuals who have been given usufruct rights to publicly owned lands for as long as they maintain satisfactory levels of production.

Forms of Urban Agriculture

It should be obvious that new technologies had to be developed and inserted into this complex panorama of production, along with extension programs in education and training and substantial supply networks for seeds, organic manure, and biological pest-control products (see chapter 4).

The GNAU recognizes four categories of lands on which urban agriculture is practiced, distinguished by their nature rather than their size. Individual Cubans are encouraged to plant *patios*, or home gardens, primarily for their own use, although they do have the right to sell or barter any surplus foodstuffs. Some 400,000 patios of the million or more that exist are registered with the GNAU, entitling the gardener access to discounted supplies, technical advice, and extension services. *Parcelas* are formerly unused parcels of land granted to an individual in usufruct as long as an acceptable level of agricultural production is maintained. These are distinguished from *huertas intensivas* and *organopónicos* in that the farmer or livestock raiser does not employ the sophisticated cultivation technologies characteristic of these two categories. By far the most important technological development that has characterized urban agriculture in Cuba is organopónicos (fig. 2.1), production units where food is grown in rectangular, walled cultivation beds (called *canteros*) roughly 1 m by 15–30 m in size, containing irrigation, special soil, and amendments. Huertas

Figure 2.1. The Organopónico Triangulo in Matanzas. (Photo by author.)

Figure 2.2. A huerta intensiva at UBPC Alamar in Habana del Este. (Photo by author.)

intensivas (intensive gardens) are similar, in that they use the same cultivation technologies as organopónicos, but at ground level rather than in
the raised canteros (fig. 2.2). This allows more flexibility in use of space,
as the size of planting beds is not specified, but huertas intensivas produce
perhaps 60% or 65% of the yield per area compared to organopónicos.

The organopónicos and the closely related huertas intensivas have
come a long way from their modest beginnings in the 1980s in HORTI-
FAR's plantain-seedling incubators.[3] By 1993, the first technical manual
on organoponic production, *Manual para Organopónicos Populares,* had
been published. It has since undergone many revisions, being frequently
updated on the basis of experience and technological innovations. The
2007 edition, titled *Manual Técnico de Organopónicos y Huertas Intensivas*
and running more than 180 pages, was produced with the participation of
several scientific institutions (GNAU 2007b: 179).[4]

The *Manual Técnico* gives precise specifications concerning the composition of the canteros in organopónico units and planting beds in huertas intensivas. In the case of organopónicos, for example, it directs that the
canteros should optimally be 1.2 m in width and between 15 and 25 m in
length—never longer than 30 m. Canteros should be positioned perpen-
dicular to the gradient of the terrain and, whenever possible, should run
north-south for best sun exposure. The substratum in which the crops
will be planted should be placed in a trench that is dug out to a depth of
30 cm. To avoid wasting space, the standard passages between the beds
should be 0.5 m wide, with wider passageways of 2 m to no more than
3 m placed between groups (*baterías*) of canteros as needed to facilitate
harvesting and other work.

The substratum itself must have various physical, chemical, and other
characteristics. Physically, it should have a high capacity for water reten-
tion, be sufficiently aerated, possess low density and high porosity, and
be structurally stable (that is, resist compaction). Chemically, it should
possess sufficient nutrients in a form that plants can assimilate easily, low
salinity, and a slow rate of decomposition. It should also be free of undesirable seed, nematodes, and other pathogens; cost little; be easy to mix;
and be resistant to external physical, chemical, or environmental changes.
The recommended mix is at least 75% by volume of organic material and
no more than 25% by volume of non-organic materials, such as soil that
does not contain high levels of clay and zeolite (a nonmetallic mineral).
The organic materials should be a combination of high-nutrient materials,

including animal manure, compost, worm humus (the nutrient-rich excreta of the worm *Eisenia foetida*), residues of sugarcane processing, materials such as rice and coffee husks produced in the processing of these crops, zeolite, sawdust, and peat. Peat, being low in nutrients, should not exceed 20% (preferably 15%) of the volume of the mixture. The nutrient value of the substratum has to be replenished yearly with application of an additional 10 kg/m^2 of organic materials. Appropriate crop rotations, plant associations, and intercalations are also to be used in order to maintain the nutritional adequacy of the substratum (GNAU 2007b: 64–69).

In addition to giving explicit guidance in the construction and preparation of the beds for planting, the *Manual Técnico* devotes extensive discussions to what to grow and how to grow it:

- Recommended varieties of vegetables that can be successfully grown under Cuban climatic conditions.
- Technical advice and instruction regarding irrigation and drainage.
- Instructions for producing compost and worm humus.
- Nonchemical (that is, biological or cultural) means for controlling plant pests and diseases.

Were it not for this degree and kind of coherent and focused centralized attention, it is unlikely that urban agriculture in Cuba would have achieved the success that it has in the Special Period.

The Oversight Structures of Urban Agriculture

The guidance mechanism and the guiding institutions for the urban agriculture effort in Cuba are complex and multilayered. By 1997, INIFAT was formally supervising urban agriculture practiced in individual parcelas, patios, organopónicos, and huertas intensivas. In addition, the national agriculture movement initiated at the First National Plenary of Organopónicos expanded its scope beyond vegetable cultivation to all other urban production, including livestock raising.

This relatively informal and disorganized movement came together under a central guiding structure in 1997 with the organization of the GNAU, headquartered at INIFAT. GNAU is a fairly compact unit of around three dozen highly qualified professionals drawn from the various research institutes and ministries involved in the urban agricultural effort. Its function is not administrative; rather, it is the highest organ in the national government responsible for cohering, mobilizing, stimulating, and promoting urban agricultural production in decentralized base units all across the urban landscape. Table 2.3 lists the various entities that participate in GNAU, as either full permanent members or collaborating organizations.

Table 2.3. Entities participating in the Grupo Nacional de Agricultura Urbana (GNAU) as members or collaborating organizations

Participating Government Ministries with Permanent Membership

Ministerio de la Agricultura (MINAG)	Agriculture Ministry
Ministerio de Azúcar (MINAZ)	Sugar Ministry
Ministerio de Educación (MINED)	Ministry of Education
Ministerio de Educación Superior (MES)	Ministry of Higher Education
Ministerio de las Fuerzas Armadas Revolucionarías (MINFAR)	Ministry of the Revolutionary Armed Forces
Ministerio del Interior (MININT)	Interior Ministry

Institutions with Permanent Membership

Instituto de Investigaciones Fundamentales en Agricultura Tropical "Alejandro de Humboldt" (INIFAT)	Alexander Humboldt Institute of Fundamental Research in Tropical Agriculture
Instituto de Fruticultura Tropical (IFT)	Institute of Tropical Fruit Culture
Instituto Nacional de Ciencias Agrícolas (INCA)	National Institute of Agricultural Sciences
Instituto de Investigaciones Avícolas	Poultry Research Institute
Instituto de Ciencia Animal (ICA)	Institute of Animal Science
Instituto de Investigaciones del Arroz	Rice Research Institute
Instituto de Suelos (IS)	Soil Institute
Instituto de Investigaciones de Riego y Drenaje (IIRD)	Institute of Research on Irrigation and Drainage
Instituto de Investigaciones en Sanidad Vegetal (INISAV)	Institute for Plant Health Research
Instituto de Medicina Veterinaria (IMV)	Veterinary Medicine Institute
Instituto de Investigaciones en Forestales, Café y Cacao	Forestry, Coffee and Cacao Research Institute
Instituto de Investigaciones Porcinas	Pork Research Institute
Centro de Investigaciones en Mejoramiento Animal (CIMA)	Center for Research on the Improvement of Animal Stock
Empresa Nacional de Ganado Menor	National Enterprise for Small Animal Stock
Empresa Nacional de Proyectos Agropecuarios (ENPA)	National Enterprise for Agricultural Projects
Centro Nacional de Producción Agropecuaria del MININT	MININT National Center for Agricultural Production

continued

Dirección de Enseñanza Técnica y Profesional del MINED	Technical and Professional Instruction Administration of MINED
Dirección de Logística del MINFAR	Logistics Administration of MINFAR
Dirección de Alimentos del MINAZ	Food Administration of MINAZ

Collaborating Nonmember Institutions

Instituto de Investigaciones Hortícolas "Liliana Dimitrova" (IIHLD)	Liliana Dimitrova Institute for Horticultural Research
Centro Nacional de Sanidad Agropecuaria (CENSA)	National Center for Plant and Animal Health
Universidad Agraria de la Habana "Fructosa Rodríguez Pérez" (UNAH)	Fructosa Rodríguez Pérez Agrarian University of Havana
Instituto Finlay–Polo Científico	Finlay Institute Scientific Hub
Universidad Central de Las Villas (UCLV)	Central University of Las Villas
Instituto de Higiene de los Alimentos del MINSAP	Food Hygiene Institute of the Ministry of Public Health
Comités de Defensa de la Revolución (CDRs)	Committees for Defense of the Revolution
Federación de Mujeres Cubanas (FMC)	Federation of Cuban Women
Asociación Nacional de Agricultores Pequeños (ANAP)	National Association of Small Farmers
Sindicato Nacional de Trabajadores Agrícolas y Forestales	National Union of Agricultural and Forestry Workers
Centro Nacional de Producción de Animales de Laboratorio	National Center for the Raising of Laboratory Animals
Asociación Cubana de Producción Animal (ACPA)	Cuban Association of Animal Production
Asociación Cubana de Técnicos Agrícolas y Forestales (ACTAF)	Cuban Association of Agricultural and Forestry Technicians
Fundación Antonio Núñez Jiménez— La Naturaleza y el Hombre	Antonio Núñez Jiménez Foundation— Nature and Man
Ministerio de Ciencia, Tecnología y Medio Ambiente (CITMA)	Ministry of Science, Technology, and the Environment
El Poder Popular a sus diferentes niveles	People's Power [government] at all different levels

Source: Companioni Concepción 2007: slides 10–11.

GNAU is at the top of a pyramid of similar committees at the provincial and local levels: there are 14 similarly tasked provincial (GPAU) and 169 municipal (GMAU) groups (Rodríguez Nodals and Companioni Concepción 2006). Each is headed by a leading official of the provincial or municipal government (for example, a municipal vice president), draws members from all involved institutions, and is guided politically by the Communist Party of Cuba. Their responsibility is to mobilize, promote,

Table 2.4. Categories and subprograms of the urban agriculture movement

Crops

Hortalizas y Condimentos Frescos	Vegetables and Fresh Condiments
Plantas Medicinales y Condimentos Secos	Medicinal Plants and Dried Herbs
Frutales	Fruit Trees
Arroz Popular	Small-Scale Rice Production
Plantas Ornamentales y Flores	Ornamental Plants and Flowers
Forestales, Café y Cacao	Forest, Coffee, and Cacao Trees
Plátano Popular	Small-Scale Banana Production
Raíces y Tubérculos Tropicales	Tropical Root Crops and Tubers
Oleaginosas	Oil-Bearing Crops
Frijoles	Beans
Maíz y Sorgo	Maize and Sorghum
Organoponía Semiprotegida	Semi-protected Organoponic Cultivation

Animal Raising

Avícola	Poultry
Cunicultura	Rabbit Raising
Porcino	Pig Raising
Ovino-Caprino	Sheep/Goat Raising
Ganado Mayor	Cattle Raising
Apícola	Beekeeping
Acuicultura	Fish Farming

Support

Control, Uso y Conservación de la Tierra	Control, Use, and Conservation of the Soil
Abonos Orgánicos	Organic Fertilizers
Semillas	Seeds
Riego y Drenaje	Irrigation and Drainage
Alimento Animal	Animal Feed
Comercialización	Marketing
Pequeña Agroindustria	Small-Scale Agricultural Industry
Capacitación	Education and Training
Integración Agroecológica	Agroecological Integration

Source: GNAU 2007b: 3.

control, and establish mandatory policy guidelines for all urban agriculture units within their sphere of governance.

As the urban agriculture program that these groups spearhead has grown and proliferated, the movement has been divided into numerous subprograms. For example, the organopónico effort (despite its importance as an impetus for urban agriculture) is part of the Vegetables and

Fresh Condiments subprogram. As of the 2008–2010 triennium, there are 28 subprograms grouped under three categories (see table 2.4): crops (with 12 subprograms); animal raising (7 subprograms); and support areas, such as technological assistance, input provision, and education and training (9 subprograms).

Certain subprograms, especially those in the input supply and technical support areas, are assigned to the supervision of technical institutions with expertise in the relevant subject. For example, the Control, Use, and Conservation of the Soil subprogram is overseen by the Empresa Nacional de Proyectos Agropecuarios (National Enterprise for Agricultural Projects) and the Instituto de Suelos (IS, Institute of Soils); similarly, the Agroecological Integration subprogram is under the authority of the Ministerio de Ciencia, Tecnología y Medio Ambiente (CITMA, Ministry of Science, Technology and the Environment) (GNAU 2007a: 7–9).

Inspection Protocols

GNAU's role in supervising the 28 subprograms can best be described as indicative planning, meaning numerical projections for the economic future. It publishes the *Lineamientos para los Subprogramas de la Agricultura Urbana* (Guidelines for the Subprograms of Urban Agriculture) every three years. For each of the 14 provinces plus the Municipio Especial de Isla de Juventud, the *Lineamientos* establishes quantified production goals in each of the 19 crop and animal-raising subprograms. It also lists other objectives, both quantitative and qualitative, for all 28 subprograms, along with verifiable indicators of success in meeting the objectives.

Perhaps GNAU's most important function is the assessment and evaluation of all urban agricultural activities, from the local level of a single production unit up to the provincial level. Members of GNAU make quarterly inspection visits (called *recorridos*) to evaluate agricultural performance, based on the indicators published in the *Lineamientos*. Every one of the 169 municipalities (*municipios*) and a substantial subset of the 1,452 popular councils (*consejos populares,* the next smaller administrative unit of Cuban territory) are visited every quarter.

This comprehensive evaluation process evolved beginning with the initial GNAU inspection visits. The evaluation methodology was submitted to the Administrative Council of Varied Crops, the Administrative

Council of the Ministry of Agriculture, and the Agro-nutritional Department of the Central Committee of the Communist Party of Cuba, and was approved with revisions in 2004.[5] The methodology so approved is applied uniformly throughout Cuba. Each year during inspection visits, every popular council, municipality, and province is rated as good, average, or poor, on the basis of separate good-average-poor evaluations of the 28 subprograms. To give a sense of the thoroughness with which GNAU approaches these evaluations, here is a summary of the process applied to the popular councils (GNAU 2007a):

- In each (quarterly) recorrido, GNAU will visit all of the councils in the "small" municipalities of Candelaria, Cerro, 10 de Octubre, Regla, Varadero, Nuevitas, Antilla, and Caimanera (most but not all of these are in Havana). In the remaining municipalities, no less than 25% of the councils will be visited in any given recorrido. Over the course of any given year *all* councils will be inspected.
- To receive a general evaluation of "good," the council must obtain a rating of good in all of the five subprograms of Vegetables and Fresh Condiments, Seeds, Organic Fertilizers, Fruit Trees, and Irrigation and Drainage, as well as in at least 80% of the subprograms evaluated in the popular council. The grade will drop to "average" if between 70% and 80% of the subprograms are evaluated as good. If fewer than 70% of the subprograms achieve an evaluation of good, the popular council will receive a "poor" rating.
- A rating of "good" requires the presence of all applicable subprograms in the popular council. In general, 27 subprograms are required, although this number may be reduced in dense urban settings such as Habana Vieja, Centro Habana, Plaza, and Cerro in Havana, or in zones with special characteristics, such as Varadero, the tourism and beach municipality; Antilla in Holguín; or Caimanera in Guantánamo.
- In each popular council, at least one-half plus one of the existing subprograms will be evaluated, including all of the aforementioned five that are required for a rating of "good." Those not evaluated in the current visit will be evaluated in subsequent visits. If a popular council is evaluated as "average" or "poor" in a recorrido, it will be visited again in the next recorrido.

- The failure to supply social needs (deliveries to schools and hospitals) adequately will disqualify the council from receiving a grade of "good."
- MINAZ has a program called the Álvaro Reynoso task, initiated in 2002, which is converting much of the sugarcane lands in Cuba to food-crop production or reforestation. For 2008–2010, failure to receive a good evaluation in carrying out the Álvaro Reynoso task will be similarly disqualifying (from an overall grade of good).

Rather more complicated procedures, including a grading system based on a 0–100-point scale, apply at the municipal and provincial levels. These scores are essentially compiled from the results achieved at the popular council level. The rigor with which these evaluation criteria are applied is evident from the fact that in the 35 recorridos between 1997 and 2006, only 6 municipalities (out of 169) received a "good" evaluation on all visits: Güira de Melena, San José de las Vegas, Yaguajay, Santa Clara, Camagüey, and Sagua de Tánamo (Rodríguez Nodals 2006: 26).

There are two other parallel evaluation processes that can grant organopónicos, huertas intensivas, or patios the meritorious statuses (or the candidacy for such status) of "municipal, provincial, or national mention" (*de referencia municipal, provincial, o nacional*) or "excellence" (*de excelencia*). Individual municipalities can receive or be nominated as *de referencia nacional* (of national mention). These designations will be discussed in greater detail in chapter 5.

To get a concrete sense of how this indicative planning and evaluation process and methodology is applied in different subprograms, we will consider three in some detail: one in crop cultivation, one in animal raising, and one in technical support.

Planning and Evaluation of the Vegetables
and Fresh Condiments Subprogram

Vegetables and Fresh Condiments (Subprogram 5) encompasses (and is organized on the basis of) organopónicos, huertas intensivas, parcelas, patios, suburban farms, and self-provisioning gardens of state enterprises (*autoconsumos*). It has been the largest and most successful of the crop-production subprograms, with a thousand-fold increase in output from

Table 2.5. Production plans for the Vegetables and Fresh Condiments subprogram in organopónicos and huertas intensivas, 2007–2010 (in metric tons)

Territory	2007 plan (t)	2008 plan (t)	2009 plan (t)	2010 plan (t)
Pinar del Río	108,066	109,494	111,636	112,304
La Habana	115,361	116,789	118,931	119,645
Ciudad de la Habana	98,000	99,428	101,570	102,284
Matanzas	103,134	104,562	106,704	107,418
Villa Clara	90,234	91,662	93,804	94,518
Cienfuegos	144,234	145,662	147,716	148,430
Sancti Spíritus	59,090	60,518	62,660	63,374
Ciego de Ávila	71,640	73,068	75,210	75,924
Camagüey	114,134	115,562	117,704	118,418
Las Tunas	67,253	68,681	70,823	71,537
Holguín	72,587	74,015	76,157	76,871
Granma	154,697	156,125	158,267	158,981
Santiago de Cuba	109,153	110,581	112,723	113,437
Guantánamo	100,517	101,853	103,995	104,709
Isla de la Juventud	4,900	5,000	5,100	5,150
Total	**1,413,000**	**1,433,000**	**1,463,000**	**1,473,000**

Source: GNAU 2007b: 25.

Table 2.6. Production goals for parcelas and patios, 2007 (in metric tons)

Territory	Parcelas			Patios		
	Number	Area (ha)	Production goal (t)	Number	Area (ha)	Production goal (t)
Pinar del Río	17,314	2,458	167,000	37,150	1,833	89,045
La Habana	8,568	3,145	270,000	58,238	1,037	60,000
Ciudad de la Habana	8,939	2,651	150,000	37,976	734	35,802
Matanzas	9,624	1,611	100,000	41,200	824	46,000
Villa Clara	5,010	2,310	180,000	20,773	1,128	35,800
Cienfuegos	13,456	583	80,000	34,500	523	66,000
Sancti Spíritus	4,723	1,456	128,055	8,312	547	22,550
Ciego de Ávila	3,359	1,966	167,000	14,015	1,497	61,879
Camagüey	9,385	1,504	129,000	15,639	1,655	93,000
Las Tunas	6,826	1,978	120,000	17,954	761	17,345
Holguín	10,266	2,182	170,000	1,241	722	58,230
Granma	15,846	1,792	117,000	17,997	607	103,805
Santiago de Cuba	12,345	1,872	160,000	38,413	4,158	58,389
Guantánamo	9,885	1,608	147,000	26,865	1,024	45,000
Isla de la Juventud	324	81	8,233	380	15	867
Total	**135,870**	**28,197**	**2,093,288**	**370,653**	**17,065**	**793,712**

Source: GNAU 2007b: 26.

1994 to 2006 (Rodríguez Nodals, Companioni Concepción, and González Bayón 2006: slide 9). GNAU's indicative planning by province, as published in early 2007, is shown in tables 2.5 and 2.6.

Objectives. The stated objectives for the Vegetables and Fresh Condiments subprogram for 2008–2010 are as follows (GNAU 2007a):

- Produce the volume of vegetables and fresh condiments specified in the plan for each year.
- Maintain no fewer than 10 different crops in each production unit throughout the year. The only exceptions are for parcelas, huertas intensivas not outfitted with micro-aspersion irrigation systems, and small house gardens less than 100 m² in area.
- Plant more than one variety of each crop in all municipalities.
- Prioritize leafy vegetables, radishes, green beans, and fresh herbs in organopónicos. In parcelas, huertas intensivas, and patios expand the range of cultivars to include staked tomatoes in the cold season, *placero* (common) tomatoes in the spring and summer, early cabbage, garlic, onions, Chinese cabbage, eggplant, and a variety of other crops.
- Pay special attention to methods of measuring production by weight.
- Verify the gross area of each production unit. Report yields on the basis of the gross area, never on the basis of the net area.
- Promote the creation and consolidation of huertas intensivas and organopónicos on MINAZ lands where sugarcane cultivation is being discontinued.
- Increase the production of fresh herbs in educational centers.
- Ensure that the Ministerio del Interior (MININT, Ministry of the Interior) consolidates production in penitentiaries and detention centers.
- Continue increasing intercropping in organopónicos, huertas intensivas, and patios.
- To achieve standardization in output measurement, ensure that vegetable or herb bunches weigh 460 g (1 lb), except for cilantro, culantro,[6] and aromatic herbs.
- Ensure that all settlements with more than 15 houses have one of the following modalities for growing vegetables and fresh

condiments: organopónico, huerta intensiva, parcela, patio, or *organoponía semiprotegida* (units covered with agrotextile shade cloths to protect plants from the intense tropical sun; see chapter 6 and fig. 6.1).

- Ensure adequate phytosanitary control in the production units, emphasizing biological controls, disinfection stations, and maintenance of the lines of production in the Centros de Reproducción de Entemopatógenos y Entomófagos (CREEs, Centers for the Reproduction of Entemopathogens and Entemophages; see chapter 4).
- Guarantee production levels for yields destined for the Ministerio de Educación (MINED, Ministry of Education), the Ministerio de Salud Pública (MINSAP, Ministry of Public Health), and selected military units.
- Guarantee that the Special Program for Increased Employment and Vegetable Production is proceeding satisfactorily, to ensure adequate conditions for the employment of women.
- Insist on a minimum of three different crops in parcelas throughout the year. It may be permissible, as a special case during the summer, to have a plot with only a single crop, with emphasis on Chinese green beans, okra, and squash.
- Promote the cultivation of rarely grown vegetables such as chayote and chicua, among others.
- Expand educational and technical training in various ways of preparing and consuming vegetables.
- Require at least five different crops throughout the year in huertas intensivas that do not have micro-aspersion irrigation systems, and three crops per year in small house gardens less than 100 m².

Evaluation criteria. The following indicators are to be used in the evaluation process:

1. Yields in kg/m²/year (see table 2.7 for criteria).
2. Fulfillment of production target: 90% or more of the production goal must be reached for a "good" rating; "average" is 81–89% of the goal; and "poor" is 80% or less.
3. Direct evaluation of crop management and level of exploitation of the productive capacity, according to the following criteria:

Table 2.7. Annual yield criteria for GNAU evaluations of urban production units (in kg/m^2)

Rating	Organopónicos	Huertas intensivas	Parcelas
Good	15–20	12–15	8–10 or more
Average	12–15	10–12	5–8
Poor	< 12	< 10	< 5

Source: GNAU 2007b: 23.

Rating of "Good"

All the following conditions are met:

- There is no weed infestation.
- There are no unutilized planting beds or canteros.
- At least 10 different crops are being grown (applies to organopónicos, huertas intensivas with micro-aspersion irrigation, and patios larger than 100 m^2).
- The level of intercropping is acceptable (at least 50%).
- More than one variety per crop is dominant at the level of the municipality.
- Plant pests and diseases are adequately controlled.
- There are no irregularities in the new units of production of the Special Plan.
- Compensation for the agricultural workers is accurately connected to the final yield produced.
- More than 90% of the available space in the unit (including the periphery) is being utilized.
- Sufficient quantities of green beans, cucumbers, and tomatoes are being grown.
- The unit has fulfilled all its commitments to supply MINED and MINSAP facilities (such as schools and hospitals).

Rating of "Average"

- Any of the preceding indicators is rated "average," up to no more than half of indicators.
- Appropriate phytosanitary control is maintained.
- The organoponía semiprotegida (shade cloth) component is rated "average" or "poor," but the unit otherwise would qualify for a "good" rating.

Table 2.8. Annual production targets for rabbit meat, 2008–2010 (in metric tons)

Territory	2008 goal (t)	2009 goal (t)	2010 goal (t)
Pinar del Río	150	157	160
La Habana	115	120	130
Ciudad de la Habana	140	147	150
Matanzas	70	75	80
Villa Clara	230	238	245
Cienfuegos	140	145	150
Sancti Spíritus	110	110	120
Ciego de Ávila	96	101	106
Camagüey	87	95	100
Las Tunas	97	102	110
Holguín	418	425	430
Granma	52	60	70
Santiago de Cuba	570	580	590
Guantánamo	80	90	100
Isla de la Juventud	37	40	45
Total	**2,392**	**2,485**	**2,586**

Source: GNAU 2007a: 67.

Rating of "Poor"

The unit fails to meet the established criteria for a "good" or "average" rating on any one indicator.

In addition, the *Lineamientos* specify that neem (*Azadirachta indica*) trees should be planted near every organopónico, so that the leaves and seeds may be used for pest control. Maize, sorghum, and insect-repellent plants such as marigold, basil, and French oregano are to be planted on the ends of every organopónico to serve as pest barriers.

Planning and Evaluation of the Rabbit-Raising Subprogram

Rabbit raising (Subprogram 20) is valuable as an efficient means of producing protein in urban conditions, in a short time frame, with a low level of inputs, and with high efficiency in conversion to protein. The subprogram has made important gains, especially for family-based home production, but quantitative targets and projections remain rather modest, as table 2.8 attests. These quantities amount to somewhat less than half a pound of rabbit meat per capita per year. The objectives are appropriately and obviously those of a small subprogram trying to grow. The *Lineamientos* list 13 objectives in addition to the quantitative meat production targets; four of these objectives and their associated indicators of success follow for illustrative purposes:

Objective. Increase the number of growers in zones authorized for rabbit raising, prioritizing family-based production.

Indicator. Increase the number of growers or breeding stock by 5%.

Objective. Continue consolidating the work of the local chapters (*órganos de base*) of the Sociedad Cubana de Cunicultores (Cuban Society of Rabbit Raisers).

Indicator. Organize all rabbit raisers in a functioning local chapter of the society.

Objective. Work to develop and consolidate food supplies for rabbits, starting with local resources recommended for this species. Work to ensure that all rabbit growers with available land plant mulberry and anauca trees and other species that can be used to feed rabbits, as well as creeping leguminous plants such as glycine or butterfly pea.

Indicator. Have an assured food supply for rabbits based on local resources.

Objective. Develop the use of alternative medicines for the prevention and cure of illnesses and parasites in rabbits.

Indicator. Maintain adequate hygiene in rabbit hutches and achieve an adequate zootechnical management of the rabbit stock.

Planning and Evaluation of the Agroecological Integration Subprogram

Agroecological Integration (Subprogram 28) was introduced into the national urban agriculture program in order to fully establish and consolidate the pattern of sustainable, locally based food production in every popular council. Its purpose is to strengthen the agroecological connections among the different subprograms and to deepen agroecological consciousness among all producers at the base level. The aim is to make urban agriculture so entrenched that future generations of Cubans will have both the option and the capability of meeting their nutritional needs by growing food in a healthy and productive environment.

Objectives. The *Lineamientos* list 10 objectives for this subprogram:

- Ensure compliance with soil, forestry, and water-use laws and encourage the largest possible number of interconnected subprograms in each production unit, municipality, and province, so as to ensure highly diversified agricultural activities in harmony with the urban and peri-urban environment.

- Ensure the implementation of all necessary measures for soil conservation and fertility and for the rational and efficient use of every square meter of land.
- Ensure that growers comply with the directives of Sanidad Vegetal (Plant Health Administration) concerning biological and botanical control of plant pests and diseases, and that they practice appropriate rotation and association of crops.
- In every unit and municipality promote the most productive varieties of crops and breeds of animals, thus contributing to increasing biodiversity.
- At all production sites, recycle all plant and animal wastes.
- Identify all species of cultivated plants and traditional animal breeds in the municipality that are in danger of disappearing, and work to save them from extinction.
- Check on the correct use of organic fertilizers and biofertilizers in all modalities of production.
- Promote training programs, circles of interest (described in chapter 4), meetings, and the like to propagate information about environmental laws and the proper care of the environment.
- Ensure an up-to-date environmental diagnostic report for all areas practicing urban agriculture.
- Insist in all units on the best possible work (and life) conditions for all workers, with special regard to the conditions for woman workers.

Evaluation. The evaluation process for this subprogram follows the general pattern of a "good" to "poor" system of rating indicators of success in meeting the specified objectives. As mentioned, CITMA has been charged with overseeing this program in all the territories. Perhaps at CITMA's insistence, among the indicators of success is the presence in the territory of a small or medium-size organopónico or huerta intensiva that functions without using conventional energy sources or that uses CREE-produced biological pest-control materials, that produces all of its own seeds, and whose workers have special material incentives tied to its productive output (GNAU 2007b).

Granjas Urbanas

So far the discussion has centered on the promotion, guidance, evaluation, supervision, and overall control of urban agricultural activity by GNAU and its provincial and municipal extensions. Yet there is also a robust and comprehensive operational, administrative, and executive side to the urban agriculture movement, which is centered on *granjas urbanas* (urban farms). The granjas urbanas were formed, beginning in 1995 in Havana, to provide government logistical and economic support to the incipient urban agriculture movement. Each municipio has at least one granja urbana, with some large and complicated municipios having more than one.

The granjas play a dual role. Administratively, the granja represents the state and is the governing institution for the urban agricultural units in its territory, which are all subordinate to it. Thus, granjas have the authority to enforce technological discipline and a unitary, cohesive organization. But they also have production units in their own right, thereby directly contributing to production. They serve to coordinate the activities of all institutions, associations, state administrative units, and other entities that play roles in urban agriculture. In short, they see to it that the GNAU guidelines for each of the 28 subprograms are carried out on the ground. Among the responsibilities of granjas are the following:

- To control and supervise all production and marketing by units in their territory.
- To ensure that all 28 subprograms receive adequate supplies of required inputs.
- To organize training for producers and extension agents.
- To maintain educational circles of interest in primary and secondary schools (in collaboration with MINED).
- To ensure schools and hospitals receive their allotted food supplies.
- To promote and consolidate neighborhood-level food production in patio gardens (in collaboration with local Comités de Defensa de la Revolución [CDRs, Committees for Defense of the Revolution]).[7]
- To maintain permanent relations with local municipal administrative councils and the presidents of the popular councils, in order to coordinate strategies for achieving the objectives in the 28

subprograms (Rodríguez Nodals, Companioni Concepción, and Herrería Martínez 2006).

The granjas urbanas also work directly with the population, buying their produce; selling them agricultural supplies; and bringing together isolated producers in order to train them, offer them services, regulate their technological and land-use practices, and make them aware of phytosanitary problems. The granjas are designed to be potentially profitable businesses, charging for the services they provide. They operate *consultorios-tiendas del agricultor* (CTAs, consultancies and stores for agriculturalists; that is, agricultural supply and extension stores for urban agriculture) in the municipios. Through the points of sale authorized by the local municipal administrative council, they market foodstuffs—both what they produce themselves and what they buy from other growers. Through the CTAs they also sell various inputs to urban agriculture—such as implements, seeds, biopesticides, organic manure, and worm humus and breeding stock for the worms themselves—as well as offering various extension and support services. As of 2006 there were 196 granjas urbanas in the 169 municipios, of which 135 were profitable. The eventual aim is for all granjas urbanas to become viable concerns that can meet their expenditures through the revenues generated by sales (Rodríguez Nodals, Companioni Concepción, and Herrería Martínez 2006).

Each granja is affiliated with the strongest state agricultural enterprise (*empresa*) present in the municipio (Rodríguez Nodals, Companioni Concepción, and Herrería Martínez 2006). These enterprises function as marketers, exporters, and service providers as well as producers themselves; they also have bank accounts in Cuban pesos as well as foreign currencies to provide the necessary financial backing.[8]

Urban Agriculture Representatives

Another key player in this decentralized motivational and administrative structure is the urban agriculture representative. MINAG typically has one urban agriculture representative in each popular council, although large and complex popular councils may have more than one representative, or two small, adjoining councils may be represented by one person. The urban agriculture representative acts to facilitate communication among producers, scientific research facilities, and state administrative bureaus. The representative's responsibilities include maintaining an up-to-date record of all agriculture in the council; generating a participatory diagnostic assessment of the council; promoting the implementation of new technologies and scientific advances; disseminating scientific and technical achievements; and both training the producers as well as learning their empirical and traditional know-how, in order to generalize this knowledge. He or she also discusses and controls the production plans of all urban agricultural units in the council, collects and verifies primary data on production, and oversees urban agricultural education programs in schools. In addition, it is up to the representative to work to integrate and consolidate the activities of all supply, service, and marketing units in the territory: CTAs, veterinary clinics, production centers for organic fertilizers, CREEs, and points of sale for produce. In short, the representative is a very important "bottom rung on the ladder" of the central state, and his or her function and role cannot be overestimated (Rodríguez Nodals, Companioni Concepción, and Herrería Martínez 2006).

The twin chains of central state oversight discussed in this chapter—first, the supervisory and evaluative chain of GNAU and its extensions to the local level, and second, the administrative, supply, and distribution chain from MINAG to empresa to granja urbana to urban agriculture representative—constitute a unique organizational structure. This structure is well suited to the task of overseeing an urban food production system that has as its guiding principle to decentralize as far as possible without losing control, and to centralize only to a degree that does not kill initiative. The Cuban urban agricultural movement's commitment to this kind of "centralized decentralization" is evident in two of the movement's mottos: "Production in the neighborhood, by the neighborhood, and for the neighborhood" and "The center of the urban agriculture movement is the producer" (Fuster Chepe 2006).

3

◇◇◇◇◇◇◇◇◇◇◇◇◇◇◇◇

Foundations in Education, Research, and Development

The development of agriculture in Cuba since 1960, and in particular of urban agriculture since 1990, can only be analyzed and understood by embedding it into the seismic shift in education, scientific research, and technological innovation in revolutionary Cuba. By the end of the twentieth century, Cuba had already established a rich history in agricultural research and technological innovation, most of it in the second half of the century under the revolutionary government's guidance. It should be noted, however, that the beginnings of this evolution extend to the early twentieth century, when Cuba became independent of Spain and began its life as a "non-sovereign republic" under the close supervision and quasi-control of the United States.

For instance, INIFAT, the organizational hub and leading institution of the current national urban agriculture movement, has its roots in an experimental research station established in 1904: the Estación Central Agronómica en Santiago de las Vegas (Central Agronomic Station in Santiago de las Vegas), located in Havana province. Although there were a few other fledgling private and underfunded governmental efforts in the following years—especially in sugarcane and tobacco research—the Central Agronomic Station remained the predominant and only continuously functioning agricultural research institution in Cuba in the first half of the twentieth century (Díaz Otero and García Capote 2006: 3; Pradas 2004).

Fledgling educational institutions capable of producing professional agronomists also date back to pre-revolutionary, and even pre-republic, days. The School of Agronomy, founded in 1900 by the U.S. military occupation under Military Order 266, was folded into the College of Letters and Sciences of the University of Havana in 1907. This would be the only option for postsecondary studies in agronomy until the establishment in

1952 of an agronomic engineering college at the Universidad Central de Las Villas (UCLV, Central University of Las Villas) in Villa Clara. UCLV, as we will see, has maintained a leading position in agricultural research since 1959, particularly in the Special Period since 1990. In the pre-revolutionary era, however, higher education of agronomists consisted mostly of theoretical studies in the classroom and involved little in the way of experimental research and development (García López and García Cueva 2006: 27–28). That was to change.

The overthrow of the Batista regime opened a new horizon in scientific investigations and gave new impetus to efforts in technological development and innovation. As early as January 15, 1959, in a speech given only 15 days after the triumph of the revolution, Fidel Castro asserted that "the future of our country has to be, of necessity, a future of men of science, of thinking men" (Castro Ruz 1960; my translation). The educational effort required to move toward this future began shortly after the revolutionaries came to power. Cuba's illiteracy rate was 23.1% in 1958. In 1961 a literacy campaign with 270,000 volunteers reduced that rate to 3.9% as 700,000 Cubans, 55% of them women, learned how to read and write (Martínez Martínez et al. 2004: 21).

Since then, Cuba has instituted near-universal primary and secondary education and has vastly expanded its network of universities and other sites of postsecondary education. In 1959 Cuba had only three universities. The University of Havana, the fount of all higher education in Cuba, established in 1728, was joined by two newcomers late in the republican era: the University of Oriente in Santiago, founded in 1947, and UCLV in Villa Clara, founded in 1952. In 1959, about 1,000 professors taught approximately 25,000 enrolled students at these three universities (MES 2009).

With the advent of the revolution, higher education immediately began to expand. The three existing universities started enrolling more students and establishing affiliated university centers in various provinces; by the early 1970s these had been consolidated into independent universities. In 1976 the Ministerio de Educación Superior (MES, Ministry of Higher Education) was created to oversee the rapidly expanding postsecondary education system, by now consisting of 27 institutions of higher learning, including many new universities (García López and García Cueva 2006: 26). By 2009 this number had risen to 65 centers of higher education (MES 2008), including 17 universities (Díaz Herryman 2009).

In the 2008–2009 academic year, 710,978 Cubans were attending institutions of higher learning (counting, in addition to the MES-affiliated universities, medical schools attached to the Ministry of Public Health, pedagogical universities attached to the Ministry of Education, and others). The so-called Universalization Program, whose goal is to bring postsecondary-level education to all Cubans through university extension centers in all the municipalities and programs of study at workplaces, enrolled more than 578,000 of these students. These figures translate into more than 63% of Cubans ages 18–24 currently furthering their education at the university level (MES 2008). Nearly 1 in 10 Cubans is a university graduate, and there are more than twice as many university graduates today as there were sixth grade finishers on January 1, 1959 (Pérez Cruz 2009).

Raising the educational level of the population was a necessary condition for even the relatively straightforward task of adopting and assimilating foreign knowledge and technology. But the university system was not sufficient to develop much domestic capacity to generate new scientific findings and technological innovations. The ability to do so became a pressing need with the U.S. blockade designed to isolate Cuba (Martínez Martínez et al. 2004: 22). The required research-and-development effort would have to extend well beyond the university system, as the following section documents.

Organizations Promoting and Disseminating Technological Innovation

To date, 80 centers and their doctoral-level staffs are employed full-time in the investigation, generation, transfer, and application of new technologies (Pérez Cruz 2009). The process of technological innovation is not, however, confined to special centers. It also occurs in institutionally supported mass movements of youth and of ordinary workers in their workplaces.

Youth Technical Brigade (BTJ)

The network of the Brigada Técnica Juvenil (BTJ, Youth Technical Brigade) is one of the earliest and most important worker movements. Founded in 1964, and with 177,000 members as of 2008 (BTJ 2009; Pérez Cruz 2009),

BTJ constitutes the scientific and technical arm of the Unión de Jóvenes Comunistas (UJC, Union of Young Communists). Its aims include the promotion of the earning of higher scientific and technical qualifications among young people (defined as up to age 35, the upper age limit for membership in UJC). It also encourages young people to search for and utilize scientific and technical information, as well as creative initiative, in solving technical and scientific problems. Other objectives are to stimulate the formation of multidisciplinary teams in the areas of invention and innovation, as well as to promote and publicly present the results obtained by team members, with a view to their actual application in production.

National Association of Innovators and Rationalizers (ANIR)

Workers over age 35 may join the Asociación Nacional de Innovadores y Racionalizadores (ANIR, National Association of Innovators and Rationalizers), founded in 1982 and also workplace-based. Most participants in ANIR (nearly 600,000 in 2008) are union members, although membership is open to all Cuban innovators, including housewives, campesinos, members of the Armed Forces, retirees, and the self-employed. The law that established ANIR defined an *innovation* as any new technical solution that is useful to the social unit in which it was introduced. It must provide a technical, economic, social, national defense, or internal order and security benefit, and must involve a change in the design or technology of production of a good or its material composition. A *rationalization,* on the other hand, is the new, useful, and effective solution of a technical or economic-organizational problem. The main guiding principles of ANIR are to encourage an innovative and innovating consciousness regarding efficiency and quality and to generalize and apply science and technology in the production of goods and services throughout the Cuban economy. The association acts as a sort of patent and registry office, registering innovations and ensuring that the entity where the innovation was first introduced recognizes and rewards its authors both materially and morally (ANIR 2009).

Forum of Science and Technology (FCT)

Both BTJ and ANIR actively participate in another institution designed to encourage technical and scientific innovation: the Fórum de Ciencia y

Técnica (FCT, Forum of Science and Technology). The work of the FCT revolves around the introduction and assessment of a wide range of technological solutions and innovations. Every two years, municipal and provincial forums are held to present the results of worker-generated innovations occurring at the local level in factories, farms, and other workplaces. These culminate in a national forum held in Havana, where the best of the local achievements are presented. By 2009, 16 such national forums had taken place. To illustrate the phenomenal growth of this movement, the first series of forums, in 1982, had 96,000 participants and 818 presentations at the local levels, with 100 of these presented at the national forum in Havana. The thirteenth forum, held in 2001, generated nearly 1 million presentations by some 1.5 million authors/coauthors, addressing more than 2 million problems at the local level. At the subsequent national forum in Havana, 529 papers were presented (Delegación Provincial del CITMA, Comisión del Fórum 2005).

Scientific Hubs

Another scientific-technological institution that emerged in the 1990s was the so-called Polos Científicos (Scientific Hubs). The idea, suggested by Fidel Castro himself and implemented in 1992, was to link the many biotechnological research and development and production entities that had been established in the 1980s to support the Cuban medical effort. The goal was to generate "closed cycles" that extended all the way from initial scientific research and discovery of a new innovation, to technological development and introduction, to actual production, and finally to marketing and post-sale patient care. Various centers of activity involved in research, development, production, and marketing would cooperate with each other without the need for additional layers of administration and bureaucracy (Rodríguez Cruz et al. 2003).

By the 2000s the hub system had grown to encompass three thematic hubs based in Havana. Relevant to urban agriculture is the Biotechnology Hub, known as the Polo Científico del Oeste (Scientific Hub of the West) because most of its affiliated institutions are clustered in West Havana. (The other two hubs are the Social Sciences and Humanities Hub and the Industrial Hub). The Biotechnology Hub now consists of 53 institutions, led by the flagship Centro de Ingeniería Genética y Biotecnología (CIGB,

Genetic Engineering and Biotechnology Center), which has played and continues to play an indispensable role in urban agricultural research and development.

In addition there are 13 provincial Polos Científicos-Productivos (Scientific-Productive Hubs), among which the ones in Villa Clara and Santiago were the first established. These focus on local problems and local capacity in their particular territories. The Pinar del Río hub (called the Polo Científico-Productivo de Vueltabajo, using the historical name of the region) focuses on improving tobacco and rice production. The Polo Vueltabajo is a network of hundreds of scientists, working in six university centers, in two experimental agricultural stations (one for rice in Los Palacios and one for tobacco in San Juan y Martínez), and in state empresas for rice, tobacco, forestry, and animal husbandry. The aim is to foster cooperation among the different actors in increasing exports (of tobacco) and substituting for imports (of rice) (Brizuela Roque 2008).

System of Science and Technological Innovation (SCIT)

All of these various strands of effort in science and technology in Cuba have been organized into a new Sistema de Ciencia y Innovación Tecnológica (SCIT, System of Science and Technological Innovation) under the administrative guidance and direction of CITMA, with the participation of other relevant state administrative authorities. SCIT has enabled the creation of spaces and mechanisms for integrating the efforts of all institutions engaged in scientific and technological innovation: the Polos Científicos, FCT, ANIR, BTJ, ACC and other scientific associations, as well as the Sindicato de Trabajadores Científicos (Science Workers Union).

In the 1990s, SCIT developed a programmatic focus on environmental and sustainability concerns, at least partly driven by the circumstances of the Special Period. This focus is reflected not only in the word *environment* in the name of the ministry overseeing SCIT (CITMA) but also in references to sustainability, the environment, or agroecology in the names of 7 of the 14 national scientific-technical programs created by 1996 (see table 3.1). SCIT priorities for 2002–2006 expanded to encompass the environment, education, food production, social concerns, health, biotechnology, information and communications technology, and

Table 3.1. National programs in the System of Science and Technological Innovation

Program number	Program name
001	Development of the Sugar Agroindustry
002	Food Production by Sustainable Means*
003	Agricultural Biotechnology*
004	Development of Biotechnological Products, Pharmaceuticals, and Green Medicine*
005	Vaccines for Humans and Animals
006	Sustainable Energy Development*
007	Sustainable Development of Mountainous Areas*
008	Animal Food Production using Biotechnological and Sustainable Means*
009	Development of Tourism
010	Cuban Society: Challenges and Prospects in Facing the 21st Century
011	Current Situation in the Cuban Economy: Challenges and Prospects
012	Current Tendencies in the World Economy and the System of International Relations
013	Global Changes and the Evolution of the Environment in Cuba*
014	Replacement Parts

Note: *The program name refers to issues of the environment or sustainability.
Source: Codorniú Pujals 1998, my translation.

energy (Quevedo Rodríguez 2006). Cuban science and technology efforts in these priority areas are organized on a program and project basis, with contractual relations between parties carrying out individual projects.

Cuban Governmental Investments in Scientific and Technical Education and Research

This massive organizational effort in science and technology benefits greatly from Cuba's consistent investment in its future through education. One outcome of this investment for the future of science and technology in Cuba is the level of achievement among Cuban primary school students. In 2008, UNESCO's Regional Bureau for Education in Latin America and the Caribbean published the findings of its Second Regional Comparative and Explanatory Study conducted in 16 countries and involving a representative sample of about 200,000 students. In the area of natural sciences, tested at the sixth grade level, students were ranked into five levels of accomplishment. These levels ranged from being able to relate scientific knowledge to daily situations that commonly occur in students' lives to abstract application of scientific knowledge (see table 3.2).

Table 3.2. Sixth grade science levels of accomplishment in UNESCO Second Regional Comparative and Explanatory Study

Level	Description
IV	Students use and transfer scientific knowledge to diverse types of situations, which requires a high degree of formalization and abstraction. They are capable of identifying the scientific knowledge involved in the problem at hand. These problems are more formally stated and may relate to aspects, dimensions, or analyses detached from the immediate setting.
III	Students explain everyday situations on the basis of scientific evidence; use simple descriptive models to interpret natural phenomena; and draw conclusions from the description of experimental activities.
II	Students apply school-acquired scientific knowledge; compare, organize, and interpret information presented in various formats (tables, charts, graphs, pictures); identify causal relations; and classify living beings according to a given criterion. They access information presented in different formats, which requires the use of much more complex skills than Level I.
I	Students relate scientific knowledge to daily situations that commonly occur in their context. They explain their immediate world based on their own experiences and observations, and establish a simple and lineal relation with previously acquired scientific knowledge. They describe concrete and simple events involving cognitive processes such as remembering, evoking, and identifying.
Below I	Students have not yet acquired the abilities required in Level I.

Source: Valdés 2008.

A full 34.73% of Cuban students performed at the highest level (see table 3.3), compared to only 3.06% of students in the next-best-performing country and an overall 10-country average of 2.46%. Roughly 65% of Cuban students scored in levels III or IV, compared with an overall average of about 14%. Whereas 5% of all Latin American students fell below level I, in Cuba this percentage was only 0.26%. Also notable is that in all countries but Cuba, males significantly outscored females and urban students significantly outscored their rural counterparts (Valdés 2008).

Thus, the Cuban science and technology system can draw on remarkably well-prepared human resources, not only in universities and research centers but in the population at large. And, through further investment

Table 3.3. Percentage of sixth grade students by science performance level by country

Country	Below Level I (%)	Level I (%)	Level II (%)	Level III (%)	Level IV (%)
Argentina	5.32	37.73	43.04	12.73	1.17
Colombia	2.62	31.68	51.09	13.59	1.02
Cuba	0.26	8.78	25.92	30.31	34.73
El Salvador	3.78	44.73	42.55	8.23	0.71
Panama	6.34	44.60	39.89	8.40	0.77
Paraguay	7.20	46.18	38.11	7.52	0.99
Peru	6.97	46.93	39.36	6.37	0.36
Dominican Republic	14.29	62.82	21.50	1.37	0.03
Uruguay	1.69	22.76	48.47	24.01	3.06
Nuevo León	2.59	30.98	47.78	16.38	2.28
All	5.18	38.72	42.24	11.40	2.46

Source: Valdés 2008.

based on deliberate governmental resource-allocation decisions, Cuba has achieved an outstanding status among Latin American and Caribbean countries. In 2000, compared to 16 other Latin American countries (including Brazil, Mexico, Argentina, Chile, Colombia, Uruguay, and Venezuela), Cuba had the highest level of expenditures in science and technology as a percentage of GDP (1.75%), the largest number of scientists and engineers working in research and development per million population (1,611), and the highest ratio of female-to-male research personnel (1.37). It was ranked second only to Brazil in percentage of GDP spent on research and development: 0.82 in Cuba versus 0.87 in Brazil (Martínez Martínez et al. 2004).

Cuba also outscored the rest of Latin America in a worldwide study of scientific and technological capacity in developing nations commissioned by the World Bank and carried out by the Rand Corporation (Wagner et al. 2001). This study analyzed countries on an index of Creation of Capabilities in Science and Technology that took the following into account:

- GDP per capita as a proxy for general infrastructure.
- Number of scientists and engineers per million inhabitants.
- Number of publications and patent applications.
- Percentage of GDP devoted to research and development.
- Number of universities and research centers per million inhabitants.

- Number of persons (adjusted) who completed higher-level studies abroad, to measure contact with outside sources of knowledge.
- Number of patents registered in the United States or Europe.

This index ranged from -1 to +1. Among the cohort of 16 Latin American and Caribbean countries, Cuba was ranked at the top, with Cuba and second-place Brazil as the only two countries with positive ratings (Martínez Martínez et al. 2004).

By 2006, Cuba was expending 0.93 percent of its GDP on research and development, and it possessed 0.8 Ph.D. or equivalent degree holders per 1,000 inhabitants (Quevedo Rodríguez 2006). Thus, by the first decade of the new millennium, Cuba had not only reached solid levels of achievement in science and technology but was also actively devoting resources to increasing its capacity in science and technology. This cannot but portend well for its future in these areas.

Education and Research Resources for Agriculture

The post-1959 societal and governmental attention to education and to scientific and technological research and development imparted an unprecedented dynamism to these Cuban sectors. Agriculture and, eventually, urban agriculture were poised to take full advantage of the possibilities engendered in this context. Starting from a meager base at the Universities of Havana and Las Villas, the study of agriculture-related fields expanded rapidly. Several major developments in this area preceded the Special Period. One was the establishment of the Ministry of Higher Education in 1976 and the subsequent proliferation of universities and university centers with agricultural research and teaching responsibilities. By the early 2000s, there were 12 universities offering a *carrera* (academic major) in agronomic engineering. Graduates earning the degree of engineer in agronomy are prepared to manage the production process in an agricultural enterprise with concern for the ecosystem, for diversified development using current agricultural production methodology and technology, and for relevant sociological strategies (of communication and extension).

Another change, this one reorganizational, occurred in 1976 when the agricultural colleges were separated from the University of Havana and reestablished at nearby San José de las Lajas, in Havana province. Thus was born what became the Universidad Agraria de la Habana "Fructosa Rodríguez Pérez" (UNAH, Fructosa Rodríguez Pérez Agrarian University of Havana). Part of the rationale for this relocation was the presence near San José de las Lajas of three national, government-sponsored agricultural research centers: the Instituto de Ciencia Animal (ICA, Institute of Animal Science), the Centro Nacional de Sanidad Agropecuaria (CENSA, National Center for Animal and Plant Health), and the Instituto Nacional de Ciencias Agrícolas (INCA, National Institute of Agricultural Sciences). Together these four institutions formed a single teaching and science complex serving the nation's agricultural development needs, especially the provisioning of Havana with foodstuffs (García López and García Cueva 2006: 30). At the same time, the University of Oriente transferred its veterinary education program to the more recently established neighboring University of Granma.

The Central University of Las Villas and the Agrarian University of Havana

UCLV and UNAH have emerged as the leading academic institutions in agricultural research and instruction. During the 1990s, as Cuba undertook a process of evaluation and refinement and improvement (*perfeccionamiento*) of its higher education in agriculture, UCLV was selected to oversee the assessment of agronomic engineering while UNAH performed the same function for veterinary medicine (García López and García Cueva 2006: 35).

The main research focus of the College of Agricultural Sciences at UCLV is sustainable production of basic grains. More than 30 researchers at the UCLV Centro de Investigaciones Agropecuarias (Center for Agricultural Research) and Instituto de Biotecnología de las Plantas (IBP, Institute of Plant Biotechnology) specialize in agamic micropropagation of plants in vitro (Facultad de Ciencias Agropecuarias 2009).[1] This work at IBP represents one of the most important technological innovations introduced by UCLV researchers. The in vitro process speeds up the propagation of plants immensely. Whereas a banana plant would naturally produce only a few suckers annually, in vitro propagation increases the number of offspring plants to 5,000–8,000. This rapid reproduction also permits the timely introduction of disease-resistant varieties, such as the

Honduran banana varieties resistant to the sigatoka negra disease that has decimated banana production worldwide. Especially important for Cuba is that it reduces seed imports that must be paid for in hard currency.

Although in vitro propagation was already a well-known technology in the world, its application in Special Period Cuba required energy- and input-saving adaptations. Rather than relying on electric grow lights, Cuban bio-factories (*bio-fábricas*) use solar energy—enabled by transparent glass and plastic structures and appropriate architectural design—in the propagation process. They have also introduced sterilization methods that avoid the use of autoclaves. The IBP started the very first of these bio-factories, based on UCLV-developed technologies that rapidly produce clones from healthy plant materials. The success of IBP's prototype led to a plan to establish 16 bio-factories all over Cuba (five of them in the province of Villa Clara) with the capacity to propagate 50 million plants per year. The eight bio-factories established to date have achieved an annual production level of about 25 million plants. For its accomplishments in this area, IBP was awarded the 2005 National Prize for Technological Innovation in the agricultural area (Avendaño 2006).

UNAH's lines of research include agroecology and sustainable development, quality of higher-level agricultural teaching, endogenous development of campesino communities, design and repair of agricultural machinery, and integrated pest management for crops. It also houses three important research centers: Centro de Estudios de Educación Superior Agraria (Center for the Study of Higher Education in Agriculture), Centro de Mecanización Agropecuaria (Center for Mechanization in Agriculture), and Centro de Desarrollo Agrario Rural (Center for Rural Agrarian Development) (CITMA 2009).

Higher Education in Agricultural Sciences

Not surprisingly, the serious, large-scale investment of resources in higher education in agricultural sciences, here conceived as including the study of plant crops and the raising of and care for farm animals, has yielded exceptional results. In the first 45 years after the revolution, Cuban institutions of higher learning graduated more than 12,000 agricultural veterinarians and more than 35,000 other agricultural specialists. By contrast, the 60 years of the pre-revolutionary republic produced only about 1,500 agricultural professionals, 850 of them in veterinary medicine (García

Table 3.4. Master's-level programs in agricultural sciences

Program	Institutions offering	Quality ratin
Agroecology and Sustainable Agriculture	UNAH	Certified
Animal Reproduction	CENSA, CIMA	Excellent
Plant Biotechnology	UCLV, Universidad de Ciego de Ávila	Certified
Preventive Veterinary Medicine	Universidad de Granma, UCLV, UNAH	Excellent
Sustainable Agriculture	UCLV, Universidad de Cienfuegos	Ratified
Animal Nutrition	Universidad de Granma	Ratified
Irrigation and Drainage Engineering	Universidad de Ciego de Ávila	Certified
Vegetable Biology	Universidad de La Habana	Excellent
Hydraulic Engineering	Instituto Superior Politécnico 'José A. Echeverría'	Excellent

Source: García López and García Cueva 2006: 53.

López and García Cueva 2006: 42). At the graduate level, master's degree programs abound, accredited and categorized into levels of quality—authorized, ratified, certified, and excellent—through a peer-review-based process. Table 3.4 lists these programs, together with the institutions housing them and the level of certification they have achieved.

At the doctoral level, nearly 1,000 of the close to 7,000 doctor of science degrees granted in Cuba up to April 2003 were in agricultural sciences, with the overwhelming majority being earned, in decreasing order of importance, in the UNAH-ICA-CENSA-INCA scientific-teaching complex, at UCLV, and at INIFAT (recall that this is the research center where the GNAU is headquartered) (García López and García Cueva 2006: 48).

The programs listed in table 3.4 reflect the new emphases of agriculture in the Special Period: sustainability, urban agriculture, and substitutes for petroleum-based agriculture. Especially noteworthy are the new master's programs in agroecology and sustainable agriculture at UNAH and in sustainable agriculture at the University of Cienfuegos and UCLV. The first of these was associated with the founding of the Centro de Estudios de Agricultura Sostenible (CEAS, Center for Sustainable Agriculture Studies). CEAS, founded in 1995, sees its role as the promotion of advanced education in agroecology. In addition to the master's program, it offers lower-level and shorter-duration diploma and short courses (discussed in chapter 4) that reach thousands of students in Cuba and abroad (García et al. 1999). The sustainable agriculture program at the University of Cienfuegos was started with input and consultation of both UNAH and UCLV and was associated with the establishment, in April 2003, of the Centro de Estudios para la Transformación Agraria Sostenible (CETAS, Center of Studies for Sustainable Agrarian Transformation) at Cienfuegos. In a fifth-year self-review of its accomplishments, CETAS points out in particular the significant technological innovations it has introduced to urban agriculture, as well as the award-winning participation of its professors in the Forum of Science and Technology all the way up to the national level (CETAS 2008).

Through the work of IBP, UCLV also inspired the establishment of the Centro de Bioplantas at the University of Ciego de Ávila, which focuses particularly on in vitro micropropagation of pineapple. As table 3.4 shows, this university is now an alternate university venue for the master's program in plant biotechnology (Rey Veitía 2006).

Government Agricultural Research Institutions

The research-and-development effort in Cuban agriculture has not been confined to academic institutions. Since the early 1960s, various research institutions have also been founded under the auspices of different ministries and other state administrative organs, with the goal of improving food production in Cuba. The earliest was the previously mentioned ICA. Next, in 1967, the Equipos de Investigación Agrícola de la Universidad de la Habana (Agricultural Research Teams at the University of Havana) were established, drawing from faculty members in a variety of fields with the necessary competence and interest. Two years later, these teams were used to constitute the National Institute of Agricultural Sciences (INCA) at the new UNAH campus in San José de las Lajas. The new location allowed the scientists more experimental sites as well as the prospect of more focused development in genetics, plant physiology, biometrics, and other fields. Also eventually finding its way to San José de las Lajas was the National Center for Animal and Plant Health (CENSA). Established in 1980, CENSA incorporated the research teams already assembled in the late 1970s in the Department for Animal Research within the Centro Nacional de Investigaciones Científicas (National Center for Scientific Research) (Díaz Otero and García Capote 2006: 10). Thus, by 1980, the three national research centers that would later form a research-teaching complex with UNAH had been situated in San José de las Lajas.

Away from Havana, the Instituto Nacional de Investigaciones de Viandas Tropicales (INIVIT, National Research Institute for Tropical Crops) has been in operation since 1967 in Santo Domingo, Villa Clara, 35 km outside Santa Clara. INIVIT's mission is to generate the scientific and technical bases for ensuring sustainability and competitiveness in the production of tropical roots and tubers; bananas; squash; papaya; and some vegetables such as tomato, cucumber, lettuce, and cabbage. In recent years it has focused most of its energy on biotechnology, in keeping with its proximity to UCLV and its prominent position in the Polo Científico de Villa Clara. It specializes in in vitro micropropagation of plants while also experimenting with genetic engineering to produce high-yield cultivars, biological control of plant pests and diseases, biotechnological development, and the use of organic manure and other biofertilizers (INIVIT 2009).

Meanwhile, the tropical agriculture institute (INIFAT) remains prominent after almost a century. Besides hosting and leading GNAU, it continues working to develop new technologies in seed production, new varieties of vegetables and plants, and the fabrication and application of biofertilizers and bio-stimulators (Pradas 2004). Havana province is also host to the Instituto de Investigaciones Hortícolas "Liliana Dimitrova" (IIHLD, Liliana Dimitrova Institute for Horticultural Research). Founded in 1971 on 109 ha of land, IIHLD researches genetic improvement of cultivars, techniques of cultivation, and seed production. Among its achievements in 2005 were new varieties of tomatoes and peppers, and a new technology for greenhouse production in the tropics (Silva 2006).

Attention to phytosanitary concerns also began fairly early in the revolutionary era. Students were sent to study abroad, especially in the Soviet Union and Bulgaria, in fields such as bacteriology, virology, pesticides, acarology, and nematology.[2] In 1967 the Centro Nacional Fitosanitario (National Phytosanitary Center) was established, followed in 1973 by the introduction of the Sistema Estatal de Sanidad Vegetal (State System of Plant Health). Local developments included provincial networks of Estaciones Territoriales de Protección de Plantas (Territorial Stations for Protection of Plants), Puntos de Frontera (Border Posts), and Laboratorios Provinciales de Sanidad Vegetal (Provincial Laboratories of Plant Health). Finally, 1977 saw the founding of the Instituto de Investigaciones de Sanidad Vegetal (INISAV, Institute for Plant Health Research), whose mission is to prevent or lessen pest-caused losses in production with the least possible risk to the environment and on a sustainable basis. INISAV offers to the network of units within the State System of Plant Health scientific, technical, and methodological support spanning the whole spectrum of concerns and problems in this area, including phytosanitary diagnostics, production and introduction of biological means of pest control, implementation of Integrated Pest-Management Systems, and monitoring of pesticide residues (Peralta García et al. 2006: 225; INISAV 2009).

The year 1986 produced what is undoubtedly the crown jewel of Cuban efforts to advance scientific research and development, namely, the establishment of the CIGB. This facility in West Havana, proposed and supported by Fidel Castro himself, represented an investment of more than U.S. $1 billion. It currently has a scientific staff of more than 500 professionals and is the central institution of the Polo Científico del Oeste. CIGB has shown considerable success in its main focus of medical research. Of interest here, however, are its successes in agricultural biotechnology, which are significant. About 13% of CIGB's Cuban and international patents are held for results obtained in agricultural research. From its inception CIGB has had a mission to research applications of biotechnology in agriculture. Shortly afterward, two provincial affiliates of CIGB were founded: one in Camagüey in 1989 and the other in Sancti Spíritus in 1990, both primarily focusing on agricultural research (Milanés León 2005; CIGB de Sancti Spíritus 2007).

Outcomes for Urban Agriculture

How these various efforts in education, research, and development lead to concrete, economically viable outcomes for urban agriculture is perhaps best illustrated through two representative examples: one is a research-and-development effort by a Ministry of Agriculture unit, the other stems from the work of CIGB de Camagüey. Interestingly, both examples come from outside Havana.

Example 1: Banana Cultivation

In 2004, the Grupo Técnico de Bio-Fábricas y Plátano (Technical Group of Bio-Factories and Banana) of the Ministry of Agriculture published a booklet titled *Tecnología del futuro: Una nueva concepción en la producción de plátano fruta y vianda en Cuba* (Technology of the Future: A New Conception in the Production of Fruit and Cooking Banana in Cuba). This report describes the group's research-and-development successes with a new technology of high-density banana cultivation adapted to conditions in Special Period Cuba (MINAG, Grupo Técnico de Bio-Fábricas y Plátano 2004). This technology doubles or triples the number of banana plants per hectare compared to traditional techniques, by reducing the distance between plants to as low as 1 m, and the area per plant to 2.5 or 3 m^2 (yielding 3,333–4,000 plants per hectare).[3] The basic outcomes of doubling or tripling fruit yields per hectare, and of sextupling seed yields, obviously result in significantly higher per-person incomes for the workers involved.

In addition, the technology has the following attributes (MINAG, Grupo Técnico de Bio-Fábricas y Plátano 2004: 12):

- Only organic fertilizers are used in feeding the plants.
- There is no use of chemical pesticides or herbicides.
- Adequate water and irrigation are available.
- Most importantly, only vitroplants (*yemas*) sown in plastic bags are used. This is absolutely necessary, because uniform height and shape of plants is crucial in achieving high yields in high-density cultivation.
- There is no replanting or propagation by suckers. All suckers are cut off at ground level as they appear.
- The area is continuously weeded, either manually or by weekly plowing using teams of oxen. All dead or diseased leaves are promptly removed from the plants.

The vitroplants are planted into appropriately spaced holes already containing 6 kg of organic fertilizer. After six months another 3 kg per plant is applied. In addition, liquid humus is applied to the foliage weekly, as both a fertilizer and a biological control agent against sigatoka negra (MINAG, Grupo Técnico de Bio-Fábricas y Plátano 2004: 8).

This technological package has several characteristics that suit it to urban agriculture. First, the technology is completely agroecological, and thus applicable in the urban environment. Second, it is well suited for small areas, such as backyards, gardens, and small parcels, provided that water and labor are available and that the technological prescriptions can

Table 3.5. Economic results of tests of the new banana cultivation technology at two farms (Empresa Minas in Camagüey and Sierra Maestra in Isla de la Juventud)

Measure	Empresa Minas, Camagüey	UBPC Sierra Maestra, Isla de la Juventud
Area sown in clone FHIA18	1 ha	3.7 ha
Number of workers	2	3
Salary, vacation pay, and Social Security costs	9,489 pesos	15,273 pesos
Monthly salary per worker	395 pesos	425 pesos
Total cost	23,683 pesos	45,927 pesos
Total income	54,808 pesos	209,679 pesos
Net income	31,125 pesos	163,752 pesos
Profit per ha	31,125 pesos	44,257 pesos
Cost per ha	23,683 pesos	12,413 pesos
Est. worker bonuses (50% of profits/worker/month)	778 pesos	2,274 pesos

Source: MINAG, Grupo Técnico de Bio-fábricas y Plátano, 2004.

be applied rigorously. And third, since sucker propagation, which is the component of traditional banana cultivation requiring the most skill and training, is completely avoided, novice urban agriculturalists can easily implement the technology.

In the 2001–2002 pilot project, which covered 4.43 ha, four clones (two each of fruit and cooking banana) and three densities were employed in the farm of a CCS member in Havana province. The results are displayed in table 3.5.

Although the booklet does not specify the distribution of profits, the current general guideline in Cuba (mentioned in the booklet) is that at least 50% of profits go to the direct producers. If this *vinculación a los resultados* (tying of incomes to results achieved) was carried out in these two places, the monthly income of the two workers at Empresa Minas would have increased from 395 to 1,173 pesos, and that of the three workers at UBPC Sierra Maestra from 425 to 2,699 pesos.

Example 2: Biological Control of Nematodes

Since the early 1990s, the group of mostly young scientists housed at the CIGB in Camagüey has been participating in various national programs, including Obtaining and Development of Biopesticides, Biofertilizers,

Bioregulators, and Natural Extracts (1 of 13 national programs proposed by CITMA in 1995) and Global Changes and the Evolution of the Cuban Environment (FCT 2008b). In this context, CIGB proposed as its contribution the development of noncontaminating biological products to control nematode infestations. Nematodes are small ringworms that attack the root systems of vegetables and other plants, weakening or killing the plants and thus reducing yields and output. Although they attack vegetables in all environments, plants grown in greenhouses are particularly susceptible.

The worldwide agroindustrial response to nematodes had been heavily focused on the use of chemical pesticides, most frequently methyl bromide fumigation, to kill the worms in the soil. This method has been severely criticized because methyl bromide significantly depletes the atmospheric ozone layer, having an effect 60 times greater than the much-maligned and already-banned chlorofluorocarbons. By the 2000s there were moves toward a global ban of methyl bromide, and most of the world turned to other chemical compounds such as Basamid (the chemical dazomet as marketed by the multinational BASF chemical corporation) and Agrocelhone (chloropicrin and 1,3-dichloropropane). These chemicals, though not harmful to the ozone layer, have all the other drawbacks of chemical pesticides: in addition to killing nematodes, they kill all sorts of bacterial flora in the soil, including beneficial species; they degrade agricultural soil quality; and they are toxic to people and animals.

Thanks to CIGB de Camagüey, Cuba chose a different path. Scientists there began to look for naturally occurring bacteria with nematode-pathogenic properties, which were known to exist. Fifteen years of development and testing that involved 158 different bacterial strains identified one strain, now commercially marketed as HeberNem, as the most effective nematicide.[4] HeberNem was subjected to extensive laboratory and field testing to determine dosages, mode and frequency of application, and necessary preparation of the soil. The soil must be completely free of clods, as otherwise substantial parts of the nematode population remain outside the range of bacterial action. It must also contain sufficient organic material, such as amino acids present in compost, manure, or worm humus. The irrigation and drainage processes must be functioning well, and the plants must be protected from uncontrolled erosion, splashing, and outside contamination, all of which affect the efficacy of the bacterial agent (CIGB de Camagüey 2007: 4).

Starting in 2002, HeberNem underwent field trials in greenhouses in five Empresa Citrícola locations in the provinces of Holguín, Ciego de Ávila, Camagüey, Cienfuegos, and La Habana. It achieved more than 90% reductions in nematode populations (Milanés León 2005). In comparative studies in adjacent beds in the same greenhouse, HeberNem outperformed Basamid in reducing nematode populations and in achieving higher crop yields and quality (that is, more of the produce was sellable for hard currency in the tourism sector) (Mena Campos et al. 2007: 52).

Besides being effective, HeberNem is also safe. It is nonsystemic, meaning it does not enter into the tissues of the plants being treated, and it is environmentally friendly, doing no harm to nontargeted flora or fauna (CIGB de Camagüey 2007: 2). In economic terms, when used in combination with the antifungal fungus *Trichoderma hazarium* A-34 in the joint control of nematodes and harmful fungi, it saves the producers—and the country—more than U.S. $400,000 for every 100 ha treated compared with use of imported chemical pesticides (Mena Campos et al. 2007: 5). Its relevance to urban agriculture is evident. HeberNem is agroecologically and environmentally friendly. Furthermore a substantial portion of Cuba's greenhouse cultivation, the primary site for HeberNem application, takes place in urban areas, including Havana city. HeberNem is continuing to gain national and international stature.

The commercial arm of CIGB, Heber Biotec, SA, markets Heber-Nem (and other biotechnology products) nationally and internationally (Acosta et al. 2007). So far, HeberNem has been granted patent rights, under three different patents, in Cuba, Australia, Ecuador, Colombia, Argentina, Brazil, Europe, Canada, Mexico, China, Vietnam, Israel, and Panama. In addition, patent applications have been filed in many other countries, including the United States, Japan, Russia, India, and Venezuela (CIGB 2008). In Cuba the application of HeberNem has become so widespread that by the end of 2008 more than 60% of all greenhouses were using it against nematodes (*Digital Granma Internacional* 2008). Further innovation has led to an eightfold increase in the efficiency of manufacturing this biopesticide, and CIGB de Camagüey is increasing its installed HeberNem production capacity fivefold in order to meet anticipated increases in domestic demand and future exports (Muñoz Pérez and Mojáiber 2009).

As we consider these two examples and note their close antecedents in earlier research and development, one conclusion is inescapable. The rapid and massive florescence of urban agriculture in 1990s and 2000s Cuba had a solid foundation in a highly educated population and in the revolutionary government's long-standing commitment to science and technology and to a vibrant research-and-development sector. Any effort mounted on less firmly anchored foundations surely would not have had such successful outcomes.

4

◇◇◇◇◇◇◇◇◇◇◇◇◇◇◇◇◇◇

Restructuring Worker Training, Preparatory Education, and Material Inputs for Urban Agriculture

Any productive system has to find a way of acquiring suitably skilled workers, raw materials, and intermediate inputs. This obvious truth took on a special significance for urban agriculture in Cuba. In all three respects, Cuba faced extraordinary challenges. Implicit in urban agriculture, under the circumstances of the Special Period, was the need to shift agricultural production to locations in or near cities and to employ agroecological technologies. The novelty of both the location of production and the technologies employed created significant needs for training or retraining of agricultural workers.

The urban and suburban residents from whom the urban agriculture workforce would have to be recruited fell into one of two categories. They were either recent migrants from rural areas who had personal, or at least familial, knowledge about agriculture, or they were city dwellers who had very remote or no rural agricultural roots. The first group was fairly substantial, but not as large proportionally as in other third-world countries where urban populations have exploded in the second half of the twentieth century due to migration from the countryside. To illustrate, between 1950 and 2008, Havana's population did not even quite double while Istanbul's and Mexico City's populations increased nearly tenfold. Moreover, the agricultural know-how rural migrants brought with them was largely conventional rather than agroecological, and thus of limited use in the new era. The city dwellers, on the other hand, might be dabblers in backyard gardening, but they were in no position to become serious agrarian producers capable of feeding a population.

In terms of educating agriculturalists, there were formidable difficulties as well. The informal education and intergenerational transmission of knowledge that take place within the family and community in traditional peasant settings had been significantly curtailed in Cuba. These modes of passing on information had become less relevant as the already small campesino sector shrank. The Agrarian Reforms of 1959 and 1963 (which nationalized large parcels of farmland) and subsequent additional state purchases of land left no more than 15% of the agricultural land in the hands of small farmers by 1990. Although farmers' children could inherit this land, the rapid spread of educational opportunities after 1959 led large numbers of them to choose other professions and leave farming altogether, as is happening throughout the world. The agroecological education system thus needed not only to introduce novel technological knowledge but also to interest young people, especially urbanites, in agriculture as their chosen profession.

In terms of materials, the new agroecological technologies being employed required specialized inputs, such as biopesticides, biofertilizers, worm humus, and steady supplies of improved and certified seeds. The members of urban agriculture cooperatives, especially the strictly urban dwellers, were scarcely up to the task of provisioning themselves with these prerequisites. Some processes, such as worm humus production and seed improvement and cultivation, would be feasible for small-scale agriculturalists to master eventually. But even bona fide, experienced peri-urban farmers could not produce pure lines of entomopathological (insect-killing) fungi that require specialized technical knowledge and a sterile environment for their propagation. The Ministry of Agriculture was still strongly oriented toward industrial agriculture, and its networks of input provision (of petrochemicals and machinery) had become irrelevant. New networks for production and distribution of inputs had to be built from the ground up. This chapter discusses these three challenges of switching from industrial to urban agriculture, and the ways Cuba attempted to overcome them.

Adult Training, Extension, and Education

This discussion of education covers all agroecology-focused efforts, regardless of whether the acquired knowledge was initially applied in urban or rural locations. Any widening of the base of agroecological knowledge among Cuban agriculturalists cannot help but redound to the benefit of urban agriculture.

Cuban Association of Agricultural and Forestry Technicians (ACTAF)

Perhaps the most important institution devoting itself to agricultural extension efforts is ACTAF, which is the professional organization of agricultural technicians with some formal education in agronomic fields. It enrolls dues-paying members who join voluntarily. By 2009 about one-fifth of those eligible to join were members (more than 22,000 members out of 103,000) (Sierra 2009e). ACTAF's statutes clearly identify support for sustainable agricultural development as a fundamental objective (ACTAF 2004). ACTAF (2009) describes its "strategic mission" as the combining of actions and resources to contribute to sustainable agricultural development. It publishes the agroecologically focused journal *Agricultura Orgánica,* the targeted audience for which is identified as agricultural technicians and anyone interested in the development of a sustainable agriculture in harmony with nature and society (Fundora Mayor 2006). ACTAF members are heavily involved with extension work in urban agricultural production units. In Havana, for example, 11 a.m. every Friday has been set aside for technical presentations, lectures, demonstrations, and training sessions, quite commonly led by an ACTAF member.[1]

A 2005 issue of *Agricultura Orgánica* spells out the tasks that provincial affiliates of ACTAF will be evaluated on during the routine inspection visits by GNAU. Six of the eight obligatory activities contain the words *capacitación* (training) or *extensión*. The affiliates are instructed to engage in training and consulting activities to ensure (1) the cultivation of at least 10 varieties of summer vegetables; (2) the installation of organopónicos on MINAZ lands no longer used in sugarcane cultivation; (3) an adequate response to climatic conditions; (4) the development, in collaboration with neighborhood CDRs, of home garden cultivation by families; (5) the development of a program of dissemination of technical information via mass media (bulletins, newspapers, television, and radio) as well as "informal" technical training; and finally (6) the facilitation of venues for the interchange of technical information necessary for the robust development of agricultural extension work (ACTAF 2005).[2]

In order to identify the principles ACTAF espouses in its extension work and to exemplify how these principles translate into practice, it is worthwhile to consider a 2006 call to action by the ACTAF affiliate in Villa Clara province, titled "Tarea: Maestros Viajeros de la ACTAF" (Task: Traveling Teachers of ACTAF). The objective of this program was to bring science and technology to local productive units in order to develop an ecological and sustainable agriculture in harmony with nature and society. The response to the question "What does the task consist of?" captures the spirit and organization of such grassroots effort. The workshop is described as a walk-through and visual inspection of the growing areas by a group of agricultural and forestry technicians, accompanied by representatives of the productive unit, to identify technical problems and possible solutions. The text emphasizes a team workshop format in which these leaders teach the growers through hands-on agricultural practice "to seek out the best, healthiest, and most sustainable alternatives. We should leave them with a profound motivation to improve every day the work they do and to increase agricultural yields, which will enable us to increase the production of foodstuffs for our people" (ACTAF, Filial Provincial Villa Clara 2006, my translation). Technical information in both written and video form is also offered.

National Association of Small Farmers (ANAP)

Another prominent Cuban institution active in training and education in agroecology is ANAP, a mass organization founded in 1961 to represent the interests of small farmers. After 1963 these were defined as those farmers whose holdings were smaller than 67 ha and who therefore remained private owners of their land; recently they have been joined by new farmers receiving usufruct rights to public lands. In the years following 1963, small farm owners were mostly organized into cooperatives: either a CPA or a CCS (see table 2.2). In the middle of the first decade of the twenty-first century, ANAP had more than 60,000 members in CPAs, 200,000 members who were CCS owners, and 100,000 who were CCS usufruct holders (ANAP 2009b).

In 1962, ANAP established the Centro Nacional de Capacitación de la ANAP "Niceto Pérez" (Niceto Pérez National Training Center of ANAP) in Güira de Melena in La Habana province. The center has expanded considerably and today incorporates a network of adjacent functioning cooperatives and farms that serve as hands-on teaching venues and as demonstration fields (ANAP 2009a). Its initial aim was to foster strong cooperatives by providing training in such areas as community organizing, management techniques, and financial management. But with the onset of the Special Period the center's attention largely turned to agroecology, as exemplified in the two-week courses the center currently offers, not only to Cubans but also to peasant-leaders in other "fraternal" countries. Three of these courses are cited as examples (ANAP 2009a, my translation): Agroecology and Sustainable Agriculture covers topics ranging from land tenancy and social and environmental impacts of industrial agriculture to design and maintenance of sustainable agricultural systems, biodiversity, and agroecological pest management; Integrated Pest Management focuses more directly on natural and biological control of pests, and how to produce the requisite biological media; finally, Agricultural Leadership addresses peasant organization and association, and structures of administration and communication, that promote development of sustainable agriculture.

With the assistance of faculty at the Center for Sustainable Agriculture Studies, the Niceto Pérez National Training Center has also developed an intensive *diplomado* (certificate of professional development) program, offered strictly to its members (García et al. 1999: 17). Diplomados are granted under the auspices of the Ministry of Higher Education to university graduates who, while pursuing their professions, elect to pursue structured professional development and continuing education. Earning a diplomado requires a minimum of 15 credits in various activities such as courses of study, practical training, and internships. It culminates with the completion of a thesis, which must be defended in front of an examining board (MES JAN 2009).

Farmer-to-Farmer Program of ANAP

Another agroecological training and extension program associated with ANAP has spread throughout Cuba since its introduction in 1997. Called De Campesino a Campesino (From Farmer to Farmer) it was inspired by a similarly named program operating in Mexico and Central America for many years. The premise is a participatory development of agroecological practices that include rescuing environmentally friendly traditions from extinction, generating new practices among campesino producers through empirical learning, and spreading new agricultural knowledge through horizontal exchanges among campesinos. In this approach, the extension agent ceases to be the expert who imposes a technological package on the farmer. Instead, he or she becomes a facilitator, giving help and support, most importantly in horizontal communications among farming peers (Sánchez and Chirino 1999). The program identifies and empowers campesino "promoters" who are willing to use and experiment with agroecological techniques on their farms, perhaps based on their own or local historical experience. Only 20% of the technical training by facilitators is theoretical; the other 80% involves the promoter in a process of learning through actual practice and of constantly reassessing results based on experience. The facilitator's primary role is to summarize successful experiences and generalize them through horizontal interactions among the local farmers via educational events, mini-workshops, sharing of experiences, field days, and demonstrations of methods and results achieved on the promoters' farms. As Sánchez and Chirino (1999: 29, my translation) put it, "There is no one more enthusiastic than a farmer who has succeeded in increasing his production with a technological innovation. Nobody will be as capable as he of encouraging his neighbor to follow his example." From its inception with a 1997 pilot project in Villa Clara that had 27 promoters carrying out 221 agroecological practices on 280 ha of land (Sánchez and Chirino 1999), the movement has spread across the length and breadth of Cuba. By the Second National Agroecology Meeting in 2006, ANAP was able to report that Farmer-to-Farmer programs were operating in 125 of the 169 municipalities in Cuba (ANAP 2006). By 2007 more than 2,000 farms and 7,000 farmworkers were involved in the Farmer-to-Farmer program in Guantánamo province, at the eastern end of Cuba far away from Villa Clara (Mejías Osorio 2007).

CEAS Certification in Agroecology

An additional important component of the Cuban training efforts in agroecology is the Center for Sustainable Agriculture Studies at UNAH. CEAS was founded in 1995 by five UNAH faculty members with a history of involvement in and knowledge of the organic agriculture movement internationally. Its general objectives are research, development, and advocacy of agroecology and the marshalling and organizing of UNAH's available resources in this area. Prominent among its efforts are training and extension activities among farmers, technicians, professionals, and administrators to enable them to master the principles of agroecology used in evaluating and achieving sustainability of organic agricultural systems (García et al. 1999: 16). The most important CEAS-led effort in this direction has been the introduction of a successful diplomado program in agroecology. Since 1996, CEAS has been awarding diplomados in agroecology in all 14 provinces through its distance learning program (Muñoz et al. 2004: 2). An experienced academic coordinator trained in CEAS workshops guides students through three courses: Agroecology, Design and Management of Sustainable Systems, and Agroecology and Sustainable Development. To receive the diplomado, each student must prepare and propose a project that could, in principle, be implemented in the institution where he or she works. In fact, many of the projects approved for implementation by CITMA originate in theses of CEAS diplomado students.

National Institute of Agricultural Sciences (INCA)

In yet another area of sustainable agricultural practice, INCA has led Cuba's participation in the Mesoamerican Participatory Plant Breeding Program, carried out simultaneously in Cuba, Mexico, and all Central American countries except for El Salvador. Since 2000 the Cuban program has allowed rural farmers access to plant gene banks maintained in the state sector or in scientific centers. The farmers can then experiment with growing different varieties and select the ones best suited to their (or their customers') particular needs and preferences. In 2003, INCA, in association with GNAU, extended seed bank access to urban agriculturalists, offering disease- and pest-resistant varieties of bananas, okra, chile, tomatoes, lettuce, and green beans.

In 2004, INCA initiated its Escuela de Agricultores Urbanos (School for Urban Farmers) program, offering integrated theoretical and practical instruction. The program began in two pilot sites—a CCS in Havana city and a granja urbana in Havana province—and focused on cultivation of green beans as the most profitable crop. In the two participating urban sites, a farmer-leader was selected and an experimental plot of land was set aside. There participants learned how to plant seeds, care for the growing plants, and apply biofertilizer, worm humus, and compost. The participant farmers themselves designed experiments to test cultivation techniques, and carried them out with the technical help of INCA personnel. Separate meetings were held in the experimental areas, then eventually the two groups joined together to assess and discuss the results, share experiences, and extend their efforts to harvesting and preservation of the crops. Each site formed a Grupo de Investigación de Agricultores Urbanos (Research Group of Urban Agriculturalists) to continue sharing information and explore marketing options (Taset Aguilar 2005).

Agroecological Beacons Program

A further impetus toward an agroecological future also took place in 1994, when the Foros Agroecológicos (Agroecological Beacons) program was initiated as a collaboration between the Grupo de Agricultura Orgánica (ACTAF's precursor) and the United Nations Development Programme (UNDP). UN funding enabled a project called Sustainable Agriculture Networking and Extension (SANE), which established demonstration farms at three selected cooperatives (CPAs) in Havana province. In each CPA, the demonstration farm ("beacon") was managed agroecologically by people who believed in the efficacy of this approach and were capable advocates and promoters. There was a strong emphasis on training and diffusion of agroecological ideas, with the active participation of all four major institutional actors already discussed in this chapter—ACTAF, ANAP, INCA, and CEAS—and other institutions.[3] Furthermore, these demonstration farms served as sites where scientists and researchers from universities and scientific centers could carry out participatory research aimed at improving agroecological practices. The results in terms of yields and profitability could be easily assessed simply by comparing the beacon's performance over time with that of the rest of the CPA (Muñoz et al. 2004). The program has since expanded (through SANE I, SANE II, and beyond) to 17 production units in 16 cooperatives and 1 state farm, located in 11 municipalities in 6 provinces. The beacons now play a significant role in the propagation of an agroecological consciousness and of organic agricultural expertise among Cuban farmers (Muñoz et al. 2004: 17).

Expo-ferias

All the efforts described thus far, albeit designed to be participatory, are guided by a central institution. A truly grassroots extension effort is the expo-ferias, locally organized at the municipality level. These fairs provide venues to present and display recent local successes in urban agriculture for public scrutiny, admiration, and, perhaps more importantly, horizontal dissemination. Expo-ferias are where exemplary patios might be singled out for public attention or where schoolchildren get to display projects demonstrating what they have learned about urban agriculture in their circles of interest (described later in this chapter). GNAU felt these fairs were important enough to merit inclusion in the criteria used to evaluate the Education and Training subprogram of urban agriculture (GNAU 2007a: 84).

GNAU Training and Extension Activities

As might be expected from its role as the leading overseer and organizer of the national urban agriculture movement, GNAU also has a considerable direct role in training and extension activities in at least three different ways:

1. *GNAU short courses.* At its INIFAT headquarters, GNAU offers short courses on various topics of interest to urban agriculturalists. In addition, INIFAT itself is a major site for research, development, and even production in plant breeding, seed improvement, and biofertilizers (García 1999).
2. *Unidades de referencia (demonstration units).* Through its recurring quarterly inspection visits to popular councils, GNAU acquires direct knowledge about the production units at the base of urban agriculture. By 2010 more than 45 such inspection tours had taken place. GNAU has used this knowledge to identify outstanding units, which progress through a hierarchy of increasingly prestigious (and difficult to obtain) designation as *unidades de referencia* at the municipal, then provincial, then national levels, and finally as national *unidades de excelencia* (units of excellence; see chapter 5 for a fuller discussion). These four categories of progressively higher status serve as potent moral and material incentives for producers. That is not their principal objective, however. The demonstration units also can and do become places others can visit and learn from. In addition, as already outstanding units, they are often sites where GNAU introduces new technologies and practices, with the hope they might have a better chance of successful adoption and subsequent generalization.

As an example, in the spring of 2007, ACTAF organized the three-day National Workshop for the Reorganization of the UB-PCs and Their Transformation into Agroecological Farms at the training school of the Empresa Frutiflora. In attendance were 15 administrators from outstanding UBPCs located all across Cuba. On one day, the participants visited UBPC Organopónico Vivero de Alamar (profiled in chapter 7), which GNAU has singled out as an unidad de excelencia. As a report on the workshop stated, "Based on the principle of following the path of those who have found success—or are getting close to it—the participants in the workshop could verify the tangible results obtained by the UBPC" (Arteaga 2007; my translation). The visit focused on practical training, an area in which the UBPC Alamar, with its exemplary program and *aula de capacitación* (training classroom), has an outstanding reputation. Such visits aim to help the participants establish and strengthen their own UBPCs to provincial-level unidades de referencia, so that they in turn can serve as sites of horizontal interchange and extension, essentially as schools for the broader community of UBPCs and their members (Arteaga 2007).

3. *Lectures and presentations during GNAU inspection visits.* In addition to inspecting and evaluating units during their quarterly visits, GNAU professionals also give lectures and presentations in their areas of technical expertise. By the completion of the thirty-fifth quarterly recorrido in 2006, GNAU inspectors had given 5,608 technical lectures attended by 238,000 urban agriculturalists (Rodríguez Nodals 2006: 27).

To illustrate the training role of GNAU inspectors it suffices to consider two days in the life of Engineer Rafael Dukesne, an experienced GNAU professional who conducted an inspection visit in the municipality of Mantua, Pinar de Río province, during March 2009. The inspections took place on the first day, when Dukesne visited various organopónicos in the municipality, the municipal center for worm humus production, the seed cultivation farm, the nursery for ornamental plants and fruit trees, and the educational and public health facilities being supplied with produce through special contracts with the inspected units. Dukesne continued with inspections of patios (i.e., home gardens) associated with different subprograms of urban agriculture, one of which, the patio of Ezequiel Herrera Rodríguez, was singled out as a unidad de referencia nacional. The second day was entirely devoted to educational activities. First, there was an expo-feria for schoolchildren to demonstrate their knowledge in each of the 28 subprograms of urban agriculture, thereby, as the local newspaper reported, "guarantee[ing] a feeling of love in future generations towards nature as the source of food" and instilling in them both the need for productivity and a respect for the environment (Baños Fernández 2009). Following the expo-feria, Dukesne gave a presentation to an audience of farmers, UBPC members, and production unit administrators that covered multiple areas of concern: the use and conservation of soil, coping with the harmful effects of droughts, appropriate crop rotation, the agroecological movement, seed selection, and the many benefits of increased plantings of neem trees (Baños Fernández 2009).

Education of Schoolchildren

I now turn to how Cuba formally educates its schoolchildren in agroecology and urban agriculture, with the reminder that education in Cuba is never completely separated from practice. For the youngest children, those of primary school age, this education takes place through *círculos de interés* (circles of interest), which operate outside the established primary school curriculum and are not mandatory for students. Urban agricultural production units are urged to establish links with nearby primary schools. Urban farmers from that unit visit the schools to share their knowledge with interested students who have joined the circle of interest. The students also visit the production unit or start an activity on the school grounds to gain hands-on, practical experience in an area they are studying, say, irrigation and drainage or medicinal plants. By 2006 there were more than 3,000 such circles of interest operating throughout the country (Rodríguez Nodals and Companioni Concepción 2006: 5). One of the goals of GNAU's Education and Training subprogram is to establish in every municipality circles of interest relevant to all the subprograms of urban agriculture (listed in table 2.4), which would total 4,732 circles nationwide, a goal that seems within reach. Correspondingly, whether circles are "optimally" functioning in all subprograms is one of the indicators by which municipalities are judged during the quarterly GNAU evaluations (GNAU 2007a: 85).

As a concrete example, consider the case of INRE-1, the first nonmilitary organopónico in Havana (fig. 1.1), and since 2002 a unidad de excelencia. INRE-1 grows enough produce to supply three day-care centers in the vicinity. Since the Cesáreo Fernández Elementary School was established in 2002, INRE-1 has also been supplying enough produce to feed each of the 150 attending students a lunch containing at least 125 g of vegetables. But INRE-1 does a lot more for Cesáreo Fernández than just provide fresh, healthful vegetables. Its workers have more or less adopted the school and taken it under their urban agricultural wings. They constructed a small organopónico on school grounds where circle of interest students engage in actual cultivation, "strictly for educational purposes" as General Sio Wong explained. In addition, INRE-1 has helped the students construct a garden called Ruta Martiana (Martí Lane) containing all of the Cuban flora José Martí cited and described in his journal (Carrión Fernández 2006: 3).

Another attempt to raise the level of knowledge about and interest in food production and agriculture among Cuban children is the annual juried arts competition called PMA en Acción, sponsored by the UN World Food Programme (Programa Mundial de Alimentos, or PMA, in Spanish) and organized every year since 1998. There are three age categories (5–8, 9–11, and 12–18 years) within which each competitor may enter up to two works, most commonly paintings or drawings, treating an aspect of food production. Reflecting where the children live and attend school, the entries tend to depict urban agriculture. The jury awards prizes to 100 or so of the best works. What sets this apart from a typical children's art contest, besides the topic, is the significance and importance that Cuban society at large attaches to it. In addition to the World Food Programme, a variety of UN and Cuban organizations support PMA en Acción (NutriNet Cuba 2009), including the Asociación Cubana de Naciones Unidas (Cuban United Nations Association), the United Nations Children's Fund (UNICEF), the Food and Agriculture Organization of the United Nations (FAO), the Programa de Desarrollo Humano Local (Program for Local Human Development) the Consejo Nacional de Casas de Cultura (National Council of Houses of Culture), the Organización de Pioneros "José Martí" (José Martí Pioneers Organization), and the National Museum of Fine Arts. The prize-winning entries are exhibited in the atrium of the National Museum of Fine Arts in Havana—the Cuban equivalent of the National Gallery of Art in the United States (fig. 4.1). The children who

Figure 4.1. A mención-winning artwork at the 2007 children's exhibition at the National Museum of Fine Arts in Havana. (Courtesy of Museo de Bellas Artes, Havana.)

participate surely absorb the message that agroecological food production in their cities and their schools is a major concern of their society.

For 15- to 17-year-olds, the Institutos Politécnicos Agropecuarios (IPAs, Agricultural Vocational Schools) offer more formal educational opportunities. These schools, designed to train middle-level agricultural technicians, existed prior to 1991, but with the arrival of the Special Period there was a major change in both their number and characteristics. Their numbers ballooned from fewer than three dozen before the Special Period to a high of 164 in 1994, eventually consolidating at 70 schools by 2009 (Santa Cruz and Mayarí 1999: 19–20). Twelve of these schools are located in Pinar del Río; seven each in Santiago, Matanzas, and Guantánamo; and only two in Havana (Ministerio de Educación de Cuba 2009). Although consolidation has reduced the number of schools, there were plans to expand matriculation in the IPAs in 2008 (Lotti 2008). In any case, the more important change was the agroecological focus injected into the IPA curriculum. This shift in focus was in fact easily introduced. For one thing, the schools are ideal locations for the integration of schooling, experimentation, and production. In addition to classroom instruction, students work in the fields every day. As part of their education, they conduct

Figure 4.2. Worm humus production at an IPA in Pinar del Río. (Photo by author.)

experiments in the areas reserved for research and experimentation. They also grow food for their own needs. This quest for self-sufficiency has led many schools to evolve into integrated farms producing both crops and livestock. In addition, the schools enter into contracts to provide food for local consumption.

The agroecological focus is reflected in the introduction of biological pest-control methods, with some schools having their own CREEs (facilities producing insect pathogens and insect eaters); producing and using biofertilizers, compost, and worm humus (fig. 4.2); and establishing organopónicos (García 1999). Between 1992 and 1996, for instance, as the number of IPAs decreased from 154 to 143, the number of schools producing worm humus increased from 41 to 112; those producing medicinal plants jumped from 2 to 102; and those with biopesticide laboratories increased from 11 (serving 30 fields) to 50 (serving 52 fields) (Funes Aguilar et al. 2002: 99).

Most of the subprograms of the national urban agricultural movement—especially Control, Use, and Conservation of the Soil; Organic Fertilizers; Seeds; and Vegetables and Fresh Condiments—have been integrated into both the experimental and production areas of the IPAs.

This has been critical to the consolidation of these subprograms and their success in meeting production goals for self-sufficiency and supplies to communities where the IPAs are located (García 1999).

Provision of Inputs

Cuba's third challenge has been to ensure the availability of the necessary material inputs for agroecological production. Some inputs, such as land and water, are constrained by geography and climate, and the best an agricultural system can do is to practice maximal conservation and appropriate utilization in order to expand as much as possible within these natural limitations. GNAU addresses these fundamental concerns through the subprograms in Control, Use and Conservation of the Soil, and Irrigation and Drainage, which focus on suitable and adequate adaptation to natural conditions.

Other material inputs into urban agricultural production are more directly dependent on human effort, being based on technological invention and scientific discovery. In the rest of this chapter I consider three important inputs—seeds, organic and biological fertilizers, and biopesticides—as well as the system of distribution through which urban farmers acquire these products.

Seeds

Although seed selection and improvement have been part of Cuban agriculture for a long time, these efforts were accelerated and deepened after the revolution. Research at institutions such as INIFAT and IIHLD has led to improved varieties. In addition, seed banks established at INIFAT and other places ensure the preservation of the genetic plant pool of Cuba (Cornide Hernández and Ortiz Pérez 2006). After the revolution, the Empresa Nacional de Semillas (ENS, National Seed Company) was founded under the auspices of MINAG. ENS has its own seed cultivation farms, has supply contracts with other farms and farmers growing seeds, and oversees the importation of seeds not yet produced in Cuba. In pre–Special Period Cuba, it served as the national source of seeds for the large-scale industrial agriculture system.

The Special Period disrupted the operations of ENS, as it did the rest of the industrial agricultural mechanism. The result, as in the rest of the agricultural system, was decentralization and a turn to agroecological methods. This decentralization of seed production went hand in hand with a strengthening of the inspection and certification process to ensure seed quality. A new law, Decreto 175, instituted in 1992, regulated the production and use of seeds and established the Sistema de Inspección y Certificación de Semillas (SICS, Inspection and Certification System for Seeds) within MINAG (Gobierno de Cuba 1992). Now, all the seeds planted in Cuba, including those for urban agriculture, undergo rigorous control and oversight to ensure they are both safe and effective (Companioni Concepción 2007: slide 27).

Under GNAU's guidance, the decentralization of seed production in urban agriculture has taken two paths. The first was the creation of *fincas municipales de semillas* (municipal seed farms). The Seeds subprogram has a goal of establishing two seed farms in each municipality, with a few exceptions.[4] Each municipality is slated to have about 10 ha devoted to seed production, divided into two farms located in different popular councils, in order to prevent possibilities of seed cross-contamination. For example, red radishes and white radishes must be grown in separate locations. These seed farms function under the strict supervision of GNAU and SICS. Their conditions of operation have to fulfill a long list of technical requirements, ranging from disinfection stations for persons and vehicles entering the farm to the nearby presence of at least two beehives to ensure pollination (GNAU 2007a: 17–18).

The second path to decentralization is the promotion of traditional seed cultivation on the agricultural plots themselves. Urban farmers who choose to grow their own seeds receive direct technical assistance from extension agents and the research institutes where new seed varieties are developed. They also receive the requisite inspection and certification services to enable them to produce, process, and conserve seeds efficiently.

This decentralization of seed production brings with it multiple benefits. First is what might be called a macro-ecological effect: a significant increase in biodiversity due to the continuous culling, selection, and cultivation of local varieties of plants. There are also direct benefits to the producers, in that seed production conforms to local preferences. The individual local producers grow for their own use the seeds for the precise crops they have chosen to cultivate. Even the municipal seed farms are responsive to needs in their municipalities, paying attention to area climatic and soil conditions and to the food preferences of local consumers. In addition, cultivating their own seeds makes urban farmers economically more independent, as they no longer have to buy seeds, reducing their costs and enhancing their economic viability (Companioni Concepción 2007: slide 24).

The scope of coverage in these local seed programs is impressive. By 2007 there were 208 municipal seed farms in the 150 or so participating municipalities (Rodríguez Nodals 2007). These farms undertake to produce sufficient seed to meet municipal needs for more than 40 varieties of the following 20 species of vegetables and fresh condiments: lettuce, carrots, red radishes, white radishes, broccoli, cauliflower, tomato, Chinese chard, cucumbers, green peppers, parsley, eggplant, green onions, turmeric, green beans, squash, okra, sugar melon, and two kinds of small, mild chiles: *ají chay* and *ají cachucha* (GNAU 2007a: 16). By 2006, quite notable production levels had already been achieved in the municipal seed farms for some crops, particularly lettuce (7.6 t), green beans (8.8 t), and Chinese chard (5.3 t) (Fresneda Buides 2006: 36). In the case of lettuce, this level of production is sufficient to supply an average of nearly 30 ha of lettuce cultivation in each of the 169 Cuban municipalities. Fresneda Buides (2006: 36) singles out squash, chay and cachucha chiles, cucumber, okra, and radish as other crops for which seed production is progressing well.

Biological and Organic Fertilizers and Pesticides

Because the challenge of providing sufficient nutrients to growing plants is paramount, it is not surprising that subprogram 2 is Organic Fertilizers (which includes biological fertilizers) (GNAU 2007a: 10). In 2005, Cuba produced and applied nearly 10 million t of organic fertilizer, about a twenty-fold increase in seven years. Although all kinds of organic materials can be used in the substratum for the canteros of urban agriculture, the supply network is focused on three modalities: compost, worm humus, and biofertilizers. In addition to fertilizers, the production and distribution of biological and botanical pesticides has also been emphasized.

Compost

Compost is the end product of the process in which some combination of organic materials, such as animal manures or plant residues from harvesting or processing crops, undergo degradation and decomposition

Figure 4.3. Worm humus production in an Interior Ministry facility in Havana. (Photo by author.)

accompanied by physical, biological, and chemical changes. The end result is a stable fertilizer free of harmful bacteria, parasite eggs, and most weed seeds—all destroyed by the high temperatures attained in the compost pile at certain points of the composting process. Scattered through the island are more than 7,000 organic fertilizer manufacturing centers (at the national and provincial levels) and microcenters (at the popular council and individual-production-unit levels). Much composting activity also goes on in the gardens and plots of individual producers (Companioni Concepción 2007: slide 16).

Worm Humus

The worm humus used in Cuba consists of the excreta of the worm *Eisenia foetida* (California red). This worm processes organic materials very rapidly and efficiently, leading to a final product so abundant in nutrients that 1 kg of worm humus can be substituted for 10–16 kg of compost in organopónico and huerta intensiva applications (GNAU 2007b: 149–79). Worm humus is well established as an organic fertilizer in the world of organic agriculture. In Cuba it was being studied and produced to some extent even prior to the Special Period, and small quantities were being produced on the island in the mid-1980s (Martínez Rodríguez 2006: 40). By 2005, worm humus manufacture had exploded to 2.7 million t (Companioni Concepción 2007: slide 16). Currently, worm humus is produced on small (home garden), medium (UBPC sideline), and large (industrial) scales (fig. 4.3). Industrial production occurs at specialized UBPCs that typically have more than 500 m² of canteros devoted to the task, and also at the organic fertilizer centers and microcenters (GNAU 2007a: 160–61). The worms can continue producing until they reach the end of their lifespan (about six or seven years) or their population expands beyond the capacity of available organic material to sustain them. At that point they can be processed into worm meal, a good source of protein for animal, and even human, consumption (GNAU 2007a: 149–50).

Biofertilizers

Biofertilizers are biological agents (bacteria and fungi) that increase the availability and absorbability of nutrients to plants, through either their roots or foliage. The most prominent agents in Cuba are the following:

- *Azotobacter chroococcum.* Bacteria of the genus *Azotobacter* were widely used in the Soviet Union for their ability to incorporate ("fix") nitrogen from the air and make it available to plants. The Cuban strains were mostly developed at INIFAT (Novo Sordo and Hernández Barrueta 2009: 26). The resulting biofertilizer is a liquid that can be applied directly to the seed and the soil or can be sprayed on the foliage of growing plants. It significantly reduces or even eliminates the need to apply nitrogen-containing petrochemical fertilizers.
- *Rhibozium* genus. The Cuban strains of this well-known nitrogen-fixing bacterium were developed at the IS and INCA.
- Fosforina. Most Cuban soils are fairly rich in phosphorus, but in forms that cannot be readily absorbed by plant roots. Fosforina is a solubilizing agent developed at the IS that "unlocks" phosphorus, making it more available to plants (Funes Aguilar et al. 2002: 174).
- Arbuscular mycorrhizal fungi. These fungi increase plant root systems, facilitating more efficient absorption of soil nutrients. INCA was the main developer of these fungi in Cuba (Medina Basso and León Díaz 2004).

Although among the objectives of subprogram 2 is "increasing local manufacture of biofertilizers, such as azotobacter, mycorrhiza, fosforina, and others" (GNAU 2007a: 10, my translation), the production of biofertilizers is not one of the indicators for evaluation. Because biofertilizer production requires a fermentation process, most still takes place at specialized facilities. By 2004, Cuba was producing sufficient biofertilizer to apply to well more than 30,000 ha of cultivated land (Medina Basso and León Díaz 2004).

Biological Pesticides

Just as important as biofertilizers for successful agroecological cultivation of crops are non-petrochemical pesticides. Even before the collapse of the Soviet Union, and especially on sugarcane lands under MINAZ control, Cuba had started introducing biological pest controls. MINAZ started creating small-scale facilities called CREEs in order to produce entomopathogenic bacteria and fungi and entomophagic and parasitoid insects. For the most part these agents were already well known and utilized in organic agriculture worldwide. What is impressive about the changes at the beginning of the Special Period relates not so much to technological innovation as to the massive supply response. By 2000, MINAG, using the experience gained by MINAZ, had established well more than 200 CREEs scattered across Cuba, as well as three large-scale industrial plants (Funes Aguilar et al. 2002: 115). These facilities are equipped to produce pure strains of local varieties of microbes (such as *Bacillus thuringensis*), fungi (such as *Beauvaria bassiana* and *Trichoderma* spp.), and insects (such as the parasitoid *Tricogramma* spp.) (Rodríguez Nodals 2007: slide 25).

The CREEs are staffed by a technically sophisticated workforce—consisting mostly of scientists, technical specialists, and IPA graduates—who are capable of maintaining sterile environments for the production of pure strains of desired organisms. There is a strict quality-control regimen within each CREE, as well as central governmental quality control via the Provincial Laboratories of Plant Health within INISAV, which also monitor manufacturing in the CREEs (Funes Aguilar et al. 2002: 115). By 1999, Cuba was producing sufficient biological pest-control organisms to serve both urban and rural areas. In fact, the nearly 1 million ha where biological controls were employed dwarfed the total area under urban cultivation (Funes Aguilar et al. 2002: 119).

Botanical Pesticides

Another alternative to petrochemicals is offered by plant extracts that are toxic to plant pests, including insects, mites, and nematodes. Interest in the many possibilities of such botanical approaches has been rising in the field of organic agriculture worldwide, and Cuba is no exception. The botanical pesticides most widely adopted in Cuba are those derived from the neem (*Azadirachta indica*) tree. Most commonly the seeds are ground to a powder. A liquid extract made from the powder can be applied directly to the plants, or the powder can be directly introduced into the soil for nematode control. Among the many advantages of neem are that it is not toxic to humans; that it works by ingestion (not contact), thus sparing beneficial insects; and that it can be cultivated and processed fairly easily and locally, allowing for decentralized adoption. Cuba has opted to manufacture neem products industrially as well, establishing neem plantations with hundreds of thousands of trees and industrial and semi-industrial processing plants, and distributing commercial neem products through a retail network (Funes Aguilar et al. 2002: 123).

The Consultancy and Store for the Agriculturist (CTA)

With urban agriculture came a growing need to distribute necessary inputs and extension services to a dispersed urban agricultural network, including home gardeners. In 1998 this led to the establishment in Havana of the first *consultorio-tienda agropecuaria* (agricultural store and consultancy). Initially, these stores were essentially limited to selling seeds. By 2001 their name had been changed to consultancy and store for the agriculturalist (CTA, or *consultorio-tienda del agricultor*) to emphasize the extension, training, and advisory services being offered (fig. 4.4). By 2006 there were 206 CTAs spread across Cuba, mostly operating under the control of the local granjas urbanas, although the 15 municipalities of Havana alone have 52 under the jurisdiction of the Empresa Provincial de Aseguramiento y Servicios (Provincial Supply Provision and Services Company) of MINAG (Hernández Pérez 2006: 13; Rodríguez Nodals 2007).

On the material supply side, the CTAs stock all the provisions discussed in this chapter: seeds and seedlings, organic fertilizers and biofertilizers, microbial and fungal biopesticides, and neem products. In addition, they

Figure 4.4. A consultorio-tienda del agricultor in Havana. Attached is a yard containing plants for sale and the ACTAF slogan, "For an ecological and sustainable agrarian development in harmony with nature and society," painted on its wall in large red letters. (Photo by author.)

sell earthenware utensils, simple manual agricultural tools, and animal feed. Furthermore, with the help of the local director of the Organic Fertilizers subprogram they coordinate the purchase by individual buyers of breeding stock of the worms used in vermiculture (GNAU 2007a: 80).

The stores are staffed by professional and technical personnel called *extensionistas de la comunidad* (community extension agents) and have available current technical information, such as up-to-date bibliographies, technical norms, pamphlets, and brochures. Customers can also obtain phytosanitary and veterinary services directly through the CTA, in cooperation with local municipal administrations. The community extension agents participate in the ongoing urban agricultural activities in their area, training and educating workers and facilitating the adoption of new technologies. They also contribute to the development of demonstration units (unidades de referencia) and of circles of interest in the schools (Hernández Pérez 2006: 13).

Through the initiatives and programs discussed in this chapter, Cuba proved itself capable of providing the necessary human and material inputs to convert a significant portion of its food production to an urban, agroecological model despite the difficult conditions of the Special Period. This effort required the introduction of completely new technologies to many workers who had little knowledge of even traditional, conventional agriculture. The education and training of the requisite personnel is in itself an impressive achievement. All the country's scientific breakthroughs and technological innovations would have mattered little, however, without its massive commitment to train and educate potential urban agriculturalists, and the concomitant, similarly massive efforts to manufacture on a large scale the new and different material inputs required for agroecological cultivation.

5

◇◇◇◇◇◇◇◇◇◇◇◇◇◇

Creating Material and Moral Incentives
to Motivate Workers

The administrators and managers of any enterprise, public or private, have an obvious interest in recruiting and retaining well-qualified workers who work efficiently and are motivated to do their best work. Individual economic actors likewise need incentives to enter a given market as sellers of their own goods and services. Employee incentives may be characterized as material (remuneration in cash or in kind) or moral (including social esteem and recognition). Clearly, if the Cuban government wanted urban agriculture to attract skilled and motivated workers, it needed to establish financial as well as non-material payoffs to agricultural work in the cities. How it did so is the subject of this chapter.

Material Incentives

An effective incentive-based remuneration scheme for individual workers in any line of work must have at least two characteristics: first, the pay should increase in correlation with effective effort, that is, results achieved in actual production; and second, the absolute level of remuneration should be high enough in real terms to make the worker's effort worthwhile. Relatively higher and upwardly flexible incomes (in the sense that workers can increase their incomes by being more productive) also serve to attract workers into a branch of activity. It is important to keep in mind that in Cuba remuneration occurs in an unusual context of a dual monetary system. Although most workers are paid in Cuban pesos (*moneda nacional*), the practice of paying a part of their wages in convertible pesos (CUCs) is becoming increasingly more common, especially in workplaces that earn CUCs in the market. This presents a considerable additional incentive for workers, because CUCs, more or less equivalent to U.S. dollars, not only have more purchasing power but, unlike the ordinary Cuban peso, are also accepted in *tiendas recaudadoras de divisas* (TRDs, literally, foreign currency collection stores, colloquially called "dollar stores"), which sell consumer goods not available for purchase elsewhere.

Historical Background

In Cuba, the issue of appropriate material incentive structures in the economy arose most acutely in the 1980s, after the implementation of Soviet-style central planning. The hallmark of this system was a centrally determined, fixed pay scale and a tendency to excessive egalitarianism that paid scant attention to the productivity of individual workers. Essentially ignored was the long-accepted Marxist distributional dictum for the post-capitalist but pre-communist, socialist phase of social organization: from each according to his abilities, to each according to his contribution—or work. These microeconomic missteps were compounded by the attempt to impose a fairly rigid and malfunctioning central planning system onto the economy. As in the Brezhnev-era Soviet Union, these underlying problems manifested in loss of efficiency in production and faltering real returns on investment.

By 1986 the Cuban leadership had acknowledged that the economy was ailing and initiated what it called the Rectification of Errors and Negative Tendencies, without specifying in detail what this reform process would constitute (Massip et al. 2001). One of the first responses to this call came from MINFAR, then headed by Raúl Castro. Gradually, more and more companies operating within MINFAR were allowed to implement a system called *perfeccionamiento empresarial* (PE, improvement/enhancement of management). Its principle was to consider each enterprise as a combination of subsystems (such as methods of administration, accounting, finance and planning, quality control, wage/incentive structure) and to adopt in each subsystem the most modern techniques of management, including ones applied in contemporary capitalism. The companies were to be converted into somewhat autonomous units with decentralized decision making in all areas, based on "economic calculation." Companies had to be self-supporting and employ strict accounting procedures, but they also had considerable flexibility in payroll management, that is, in choosing workers and setting their salaries. The primary aim, of course, was to increase production and efficiency, but it was also hoped that the new management style would engender greater "communist" consciousness among workers (Ibáñez López 2006). By the end of 1989, PE was in place in about 10 MINFAR companies, with quite encouraging results. Overall labor productivity went up by 30% while the number of employees on payroll declined by 16% (Ibáñez López 2006). Over the next decade, MINFAR gradually extended this new system of management to most of the enterprises under its control. Impressed by MINFAR's accomplishments, the Fifth Congress of the Communist Party of Cuba, in its Economic Resolution, decided to apply PE in all state enterprises. This was formally accomplished by Decreto-Ley 187, passed in 1998.

Appropriate material incentive systems are a crucial component of PE. Among the 17 general principles of PE listed in a 2000 issue of *Cuba Socialista* are these two concerning wages and incentives:

1. Workers and administrators will be paid according to the socialist principle "from each according to his abilities, to each according to his work."
2. Collective incentives will be based on the efficiency achieved by the company and its contribution to the national economy (combined with moral recognition). Results will be rewarded, and *not* just effort. (Tristá Arbesú 2000; my translation)

By 2007 about 30% of all state enterprises were applying PE, despite some continuing difficulties—especially in maintaining adequate accounting systems—that led to occasional suspensions. In general, the application of PE has led to higher salaries, profits, and efficiency. In the first half of 2007, for example, companies practicing PE accounted for only 20% of all sales but garnered 51% of all profits and 72% of all hard-currency earnings; in addition, their productivity levels were 50% higher than in companies not yet applying PE (Lage Dávila 2007).

The principle of *pago por resultados* (pay according to achieved results) is now well established in Cuba, but, as practiced within PE, it remains a rather flawed system. If the company makes a profit, all workers receive the same bonus, regardless of individual contribution and effort. Thus, incentive payments are individually distributed, yet they remain too collective in nature, undoubtedly giving rise to problems with "free-rider" employees. A recent revision of the pago por resultados system attempts to lessen if not resolve this problem by aligning, however imperfectly, individual wages with individual contribution to output as measured by concrete indicators.[1] This regulation has no a priori, fixed upper limits to earnings by direct producers, such as urban agriculturalists. The more one produces, the higher the pay.

The pago por resultados principle is, of course, applicable in all lines of economic activity, not just in state enterprises or as part of PE. In fact, Raúl Castro gave one of the earliest and most forceful articulations of this principle in connection with agriculture. In a speech delivered at HORTI-FAR in 1997 on the tenth anniversary of the invention of organopónicos, Castro referred to his recent visits to Vietnam and China. With 22–23% of the world's population but only 7% of its arable land, China had to unleash productive forces in its agricultural sector in order to feed its population. He concluded that the same had to happen in Cuba: "the knots that bind the development of productive forces have to be undone; and that means he who works more, earns more and lives better; and our egalitarianism has to be the equality of rights. We all have the same rights, but all the rest depends on our efforts, the personal efforts of each to study, to better himself, to work; and that he who works more, contributes more, may live better; that is the true equality. The other supposed equality that we used to have and still have, that is inequality" (Castro Ruz 1997: 36; my translation).

Material Incentives in Urban Agriculture

In short, Cuba has been wrestling with the issue of how to structure material incentives for its workforce throughout the half century of its existence as a socialist republic. The issue has been acutely on the agenda since 1985, predating the era of urban agriculture. But as urban agriculture developed into a formidable branch of production in the 1990s, dealing with incentives was unavoidable. The new movement was, however, well positioned for this task. From the very start it was conceived and organized with an almost single-minded emphasis on decentralization in production, commercialization, and acquisition of inputs (Fuster Chepe 2006). In such a context, it is easy to isolate localized, and even individual, economic results, particularly so among individual *parceleros* (farmers with usufruct or other access to land) and home gardeners. In these situations the individual producer receives all profits, so pago por resultados is automatic.

The remainder of the urban agriculture workforce is divided between the cooperative and state sectors. The cooperative sector consists almost entirely of UBPCs and CCSs, so I will focus on these types. Urban UBPCs differ somewhat from most rural UBPCs in the provenance of their lands and the composition of their workforces. Because rural UBPCs were established on extant state farms, they inherited the existing workforce as their members, regardless of their productivity. Urban UBPCs, in contrast, generally obtained usufruct rights to urban tracts not previously used for agriculture and recruited their membership from the general population, giving them more flexibility in their workforces. All UBPCs, however, are governed by the law that established them, Decreto-Ley 142 of September 1993. The preamble to this law lists two important principles that should guide its implementation (Valdés Paz 1997: 243; my translation):

• A stable connection between each worker and a given piece of land so as to encourage his interest in his work and his concrete sense of individual and collective responsibility.
• The strict association of workers' incomes with the achieved results of production.

Accordingly, Ministry of Agriculture regulations allow up to 50% of profits to be distributed to individual cooperative members as income, in a manner decided by the UBPC membership (Figueroa Albelo et al. 2006).

In the CCSs, whose members are either individual owners of farmland or individual usufruct holders, the producers' profits are automatically tied to their harvests. Most CCSs also have collective organopónicos or huertas intensivas whose workers, as distinct from the parceleros or owner-members, share in any profits gained.

The remainder of agricultural production takes place in surviving state enterprises. But here, too, the principle of pago por resultados is well established, and few workers get a fixed salary independent of their production. As an example, the Empresa de Cultivos Varios Habana in Havana was severely battered in the Special Period. In response, it was restructured into 453 fincas estatales (state farms), each headed by one worker (the jefe de finca), who typically lives on the farm with his family. These "family farms" are under contract with the empresa to produce a specified product according to the overall operating plan, but they are allowed to grow other products for sale or for farmworkers' personal consumption. The farmworkers retain up to 70% of any profits to use as they wish, thus tying their incomes to their individual contributions.

The results speak for themselves. In 1993–1994 the empresa had 2,247 workers; produced 112,000 quintals of fruits, vegetables, and tubers; lost 10 million pesos; and had 335 ha under cultivation. By 2005 the empresa had stabilized its workforce at about 2,000 workers, produced 1,126,000 quintals of produce, and made 1,350,000 pesos in profit with 4,351 ha under cultivation. This workforce operated not only the 453 fincas but also 6 organopónicos, 132 greenhouses, 3 plants processing and packaging goods for export, a CREE, and a facility for drying medicinal plants.

In this context of individual responsibility and reward, the issue of state versus usufruct or private landownership does not seem particularly important (Sánchez Balboa 2006: 10). Thus, the connection of compensation to results is more or less successfully implanted in the urban agriculture sector. As champions of urban agriculture in Cuba have stressed repeatedly, however, farmworkers' incomes also have to be high enough that they can rightfully feel they are being not only justly compensated but at least relatively well paid. ACTAF President Eugenio Fuster Chepe (2006: 6) lists adequate economic incentives for the producers among the six underlying concepts of urban agriculture. GNAU President Adolfo Rodríguez Nodals and Executive Secretary Nelso Companioni Concepción (2006: 6) similarly argue that for the future of Cuban urban agriculture to be secure, its economic efficiency must be consolidated so as to assure the producers of good incomes.

Farmworkers' incomes are fundamentally generated by the revenue obtained from the sale of their products. Since revenue equals quantity multiplied by price, both factors are equally determinative in fixing income level. The quantity produced is pretty directly correlated with technology and efficient use of resources. In contrast, pricing in Cuba, especially for food, presents a complex panorama. To understand how urban agriculture fits in, one must first have an idea of the overall picture. Prior to the Special Period, except for a brief period of experimentation in 1980–1986 with a *mercado libre campesino* (peasant free market), Cubans obtained whatever food they did not produce in their own families or purchase in restaurants either legally through la libreta or, illegally, through the hard-currency *mercados diplomáticos* (reserved for diplomats and other foreign residents of Cuba) and on the black market. The Special Period brought with it monetary duality as U.S. dollars were allowed to circulate freely (although eventually replaced by the dollar-equivalent CUC peso). In addition, all Cubans with dollars (and later, CUCs) were allowed access to dollar stores, successors to the *mercados diplomáticos*. Other reforms, in 1994, led to the reestablishment of *mercados agropecuarios libres* (MALs, free markets for agricultural goods), organized by the Ministerio de Comercio Interior (Ministry of Domestic Trade). In them state, cooperative, or private sellers can offer many foods for sale at prices denominated in Cuban pesos and determined "freely" by supply-and-demand conditions.[2] Urban agriculture productive units also sell directly to the public at curbside stands near their fields and at *puntos de venta* (points of sale) in the vicinity, setting their own prices in Cuban pesos. Furthermore, home gardeners are encouraged to sell their excess production in their neighborhoods.

Not all food is sold in these high-priced environments, however. In an attempt to lower prices, the state withdrew its participation in the MALs in 1998, establishing instead its own network of state markets and points of sale under the supervision of MINAG. These *mercados agropecuarios estatales* (MAEs, state markets for agricultural goods) operate in Cuban pesos with *precios topados* (capped prices) that undercut prices in the MALs. MINAG acquires the produce and meat it sells at the MAEs via contracts with various growers (UBPCs, CPAs, CCSs, and other state and private producers). There are nearly 2,500 such outlets in Cuba (Nova González 2006: 122) and more than 300 in Havana (Cabrera Balbi 2009). The prices at urban agriculture points of sale and MAEs are typically 10%–30% lower than those charged at the MALs; although significantly lower, they still fluctuate according to the free-market, supply-and-demand-determined prices at MALs. If they did not, significant opportunities would arise for arbitrage between the two markets and profiteering by middlemen.

Other lower-priced options for consumers include markets operated by the Ejército Juvenil del Trabajo (EJT, Work Army of the Youth). This institution allows young people to fulfill their military-service obligation by working on rural farms that grow food mostly for the armed forces. Part of their output is sold directly to the public in markets organized and run by the EJT in the cities. Also, there are government-organized monthly *ferias* (market fairs) where rural growers truck in produce from the countryside to offer it for sale, mostly directly from the trucks, at prices considerably lower than at the agro-markets or the points of sale of urban agriculture. These two lower-priced alternatives, however, cannot entirely substitute for the other markets. For one thing, they are temporally and spatially few and far between. Also, these outlets, especially the ferias, typically do not offer the assortment of fresh vegetables grown in urban agriculture, concentrating instead on fruits such as bananas and pineapples, plantains, root crops, squash, rice, and beans. Third, these products, imported into the city from the countryside, are not typically guaranteed to be organically grown. Finally, in the least-profitable sales environment for urban agriculturalists, agricultural base units enter into contracts, typically at below-market prices, to supply vegetables for public agencies (schools, universities, day-care centers) and social-service facilities (hospitals, old age homes, boarding schools).

This general assessment of pricing behavior is consistent with anecdotal evidence I accumulated during my 2007 stay in Havana. The prices at the points of sale and other markets are clearly posted. In addition, the prices at the MAEs are fixed for a month at a time, and price lists are published in *La Tribuna de la Habana,* Havana's weekly newspaper. Prices in Ciudad de la Habana are considerably higher than in other provinces (Nova González 2006: 65). After all, this is the richest and most populous province, it receives disproportionate amounts of remittances from abroad, and it has the smallest per capita production capacity. With these caveats in mind, some comparative data gathered in the first four months of 2007 follow; all prices are in Cuban pesos:

Banana. Price per *mano* (literally "hand," a stem of usually 16 bananas) = 8 (MAL), 6 (point of sale, hereafter POS), 1.5–2.5 (feria)

Plantain (plátano burro). Price per mano = 8 (MAL), 5 (POS); price per pound = 0.8 (POS), 0.8 (MAE), 0.6 (EJT), 0.6 (feria)

Green peppers. Price per pound = 7 (MAL), 3–4 (POS), 2.5 (EJT)

Guava. Price per pound = 5 (MAL), 2.5 (POS), 2.5 (MAE)

Tomato. Price per pound = 2.5–12 (MAL), 3 (MAE), 2.5 (feria), 1–3 (POS)

Lettuce. Price per head = 5 (MAL), 3 (POS)

Eggplant. Price per each = 3–5 (MAL), 1 (POS). (The eggplants were much larger at the MAL than at the POS.)

Cucumbers. Price per pound = 3–8 (MAL), 3 (POS), 1.5 (MAE)

Given this complex environment, urban agricultural revenues, and thus the incomes of individual farmworkers, depend on both the market prices determined in the MALs and the proportion of goods sold in low-priced (schools, hospitals, to the state by contract) versus high-priced (directly to the public) markets. Generally, revenues and farmworker incomes are quite high because more than 70% of urban agricultural output is marketed to the public. Another 20% is directly consumed by the producers themselves, and less than 3% goes to public agencies at low contracted prices (Companioni Concepción 2007: slide 31). MAL prices—and the associated, somewhat lower prices at MAEs—have remained quite high relative to urban incomes, although they did decline significantly from pre-1994 black market prices. Given that the average monthly salary of urban dwellers is about 400 pesos, the reported average price of 1.1 peso per pound for agricultural goods (excluding meat products) seems outrageous (Nova González 2006: 312)—let alone the 12 pesos per pound for tomatoes I observed in a Havana MAL in 2007. Unfortunately, rural producers often profit little from these high prices, because a good portion of their proceeds is siphoned off by intermediaries. Urban agriculturalists benefit from being under the "umbrella" of the prices generated in the MALs and from being less dependent on intermediaries.

How does the Cuban food market support such high clearing prices? The answer would seem to lie in peculiarities in both the supply and demand sides. On the supply side, especially after the state withdrew from the MALs, sellers seem to have formed informal cartels that have given them considerable monopoly power. Their quasi-monopoly is strengthened by lack of competition from the dollar stores, whose prices are kept extremely high as an implicit tax on hard-currency earnings (which are largely received as remittances from the United States).

On the demand side, there are several explanations for why demand persists despite these exorbitant prices. First, average Cubans obtain most of their daily diet (at 3,300 kcal per day in 2005) at no cost or very cheaply through la libreta, home gardens or autoconsumo gardens at their workplaces, and meals provided at workplaces, schools, or public agencies. Nova González (2006: 283) estimates that Cubans purchase only about 9% of their calories, 16% of their protein, and 28% of their fat intake at a MAL, MAE, or point of sale. For a family of four, these purchases on the open market total about 450 pesos, or about 56% of the 800-peso average monthly income of a two-income family. Rationed products cost another

150 pesos, bringing total food expenditures to about 75% of monthly income (Nova González 2006: 124–32).

Although this percentage seems extremely high, one has to remember that education and health care in Cuba are free to consumers. Cultural goods such as books and entertainment are highly subsidized. About 85% of all Cubans own their own homes and pay no rent or property taxes. Those who do rent are on long-term leases and pay only 10% of their income monthly for housing. Utilities such as telephone service, gas, electricity, and water are heavily subsidized. Thus, the proportion of a Cuban household's income spent on living expenses is probably not all that different from that for a U.S. household. In Cuba, however, most discretionary income is (because it can be) spent on food. This concentration of spending in the MAL and urban agriculture markets helps sustain the high prices.

The urban agricultural experiment was conceived as an economically attractive option for urban activity. So far, it has indeed proved to be so. One cannot foresee what future changes will take place in the system of prices, rationing, and subsidies, but as of this writing at any rate, urban agriculture has managed to carve out a fairly lucrative space for itself in the economy. The fact that, in the 15 years since its inception, it attracted a growing workforce—more than 350,000 workers, or 7% of the entire Cuban workforce—stands as powerful evidence for this.

Moral Incentives

Because humans do not live by bread alone, non-material incentives have also played a significant role in mobilizing the urban agriculture effort. In revolutionary Cuba, such incentives are strongly cast in moral terms. The issue of moral incentives in the economy was introduced into Cuban political discourse shortly after the triumph of the revolution by Ché Guevara himself. In a 1965 essay, Guevara argued that the construction of socialism also required the constitution of a socialist "new man" with a new consciousness. This new man or woman would seek rewards beyond individual material satisfaction. In this new context, the prize would be a new society in which individuals do not sell their labor to capitalists in order to satisfy their "animal" necessities, but rather see work as a social duty and as self-realization. As Guevara put it, "The prize is a new society in which individuals will have different characteristics: the society of communist human beings" (2003: 218). For the new man, "it is not a matter of how many kilograms of meat one has to eat, or of how many times a year someone can go to the beach, or how many pretty things from abroad you might be able to buy with present-day wages. It is a matter of making the individual feel more complete, with much more inner wealth and much more responsibility" (2003: 225). Of course, Guevara was aware of the difficulties inherent in molding such a new human, which could only be done through extensive direct and indirect education under the direction of the Communist Party and the state. Material incentives, especially *collective* ones—would continue to play an important role in the transitional phase.

In the 1960s this philosophy gave rise to polemics and controversies among Cuban leaders. By the 1970s the consensus had swung decisively in the direction of emphasizing material (including collective material) incentives for workers (Zimbalist et al. 1989: 368–70). However, the concept of moral incentives, which presuppose what in microeconomic terms could be called altruistic preferences, has survived in both Cuban discourse and practice. Urban agriculturalists, for example, are urged, and sometimes induced by policy, to take overall social conditions into account in their economic behavior rather than being pure income maximizers. The provisioning of schools, hospitals, and military units with produce at below-market prices is the most visible example of this.

Of course, a lot of this "altruistic" activity is "guided" by central and local government authorities. But at least some echoes of Ché's new man exist within urban agriculture, as evidenced in the words of the administrator of the successful El Mango UBPC in the San Cristóbal municipality of Pinar del Río, which started out as a pig-raising UBPC in 1998 with 12 workers: "The secret for obtaining good results consists in combining the personal interest with the social. If we were only interested in profits, we would have specialized in pork production, and today we would still be 12 workers with high incomes, but we wouldn't have given others the possibility of a dignified employment, nor developed other lines of production necessary to the population" (Suárez Rivas and Suárez Ramos 2009: n.p., my translation). By 2007 the El Mango cooperative had grown to include 192 workers and multiple fincas and organopónicos.

Still, the existing "moral" incentives in urban agriculture by and large do not depend on the existence of the "new man." Most are structured as awards and recognition given to individuals or collectives, as well as to entire popular councils, municipalities, and provinces, for their urban agricultural achievements. This recognition of urban farmworkers takes various forms. First, there is formal recognition accorded by different levels of the national urban agriculture movement and the government. The GNAU-led inspections and evaluations of productive units (discussed in chapter 2) contain a grading system that leads to recognition for outstanding performance, analogous, say, to the Dean's List and Latin Honors systems of U.S. colleges. Second, the units that are so singled out become centers of introduction and diffusion of new technologies, are held up as examples worthy of emulation, and are visited by urban agriculturalists who want to learn from their successes. They become sites of conferences, workshops, and other extension activities, thereby receiving considerable social recognition and esteem. Third, there are individually focused mixtures of in-kind material and moral benefits and rewards in programs under the umbrella of what is called *atención al hombre* in Cuba—the equivalent of human resources in the United States. These include improvements and enhancements of the working conditions for cooperative members, ranging from the provision of wholesome, well-prepared lunches—often using vegetables grown on the premises—to well-equipped sanitary facilities, especially for women workers. Members of cooperatives are encouraged to attend classes at the municipal university extension centers in order to complete their university educations. In one UBPC I visited in 2007, about one-third of the nearly 150 members had degrees from universities or technical schools. Of the rest, seven were taking university courses in the neighborhood.

The aim of these kinds of programs is to build both self-esteem and societal esteem for urban agriculturalists. Gone is the stereotype of a poorly educated peasant who labors under the sun from sunup to sundown to eke out a meager living. It is being replaced with the image of a cultured, intelligent, scientifically and technically well-educated urban professional. Before long, urban agricultural workers are likely to be as well respected and honored by the population as are the teachers and medical personnel that Cuba has produced in such profusion and to such good effect. This prospect itself must serve as a powerful stimulant in both eliciting effort from current urban agriculturalists and attracting urban youth to a career in urban agriculture.

The bedrock of the moral incentive structure are the institutional, governmental processes for recognizing outstanding performance. Most of them take place under the auspices of GNAU, but in some cases they reach higher in the government to ministerial and even presidential levels. The most significant recognition process organized and carried out by GNAU is the referencia system mentioned in chapter 4. Production units at the base level (*unidades de base*) may aspire to receive the designations *de referencia municipal* (of municipal reference), *de referencia provincial* (of provincial reference), or, at the national level, candidacy for or designation as *de referencia nacional* (of national reference). All four levels of honors can be bestowed either on individuals (patios, parcelas) or collectively on production units such as organopónicos, UBPCs, huertas intensivas, seedling houses, or nurseries. In addition, municipalities can aspire to candidacy for or designation as de referencia nacional status, based on the overall excellence of all productive units in that jurisdiction.

The subordinate units of GNAU at municipal (GMAU) and provincial (GPAU) levels have the authority to designate base units as having referencia status at their respective levels, with the proviso that a unit must have had the status of referencia municipal for at least six months before it can be considered for referencia provincial status. After eight months of provincial status, the unit may be elevated to candidacy for referencia nacional status, awarded by GNAU. The transition to full referencia nacional status may follow after six months as a candidate, if approved by the president of GNAU, either on the recommendation of a GNAU member or on the president's own initiative.

The referencia status must be confirmed and ratified twice a year at the municipal and provincial levels according to procedures set by the local GMAU or GPAU chapter. The national reference status or candidacy for it is verified by GNAU during the third quarterly visitation each fall. Any referencia unit that receives an evaluation of "average" at any point loses its status and must reenter the referencia system at the lowest level, if after a six-month waiting period, it again meets all requisite conditions (GNAU 2007a: 91).

Besides outlining the process of promotion from lower to higher levels of referencia, the GNAU *Lineamientos* also specify, for organopónicos and huertas intensivas, the required productivity levels, economic success, and other conditions for selection as a unit of referencia municipal (GNAU 2007a: 97):

- The unit must have been in existence for at least six months, starting from the date of sowing of the first crop.
- It must have an annual output of at least 10 kg/m² in organopónicos and 7.5 kg/m² in huertas intensivas.
- The unit must be profitable, and thus able to distribute part of its profits among workers.
- It must comply with GNAU performance parameters: full use of arable area, more than 50% intercropping, free of weeds, adequate levels of organic material in the beds and substratum, adequate protection of the irrigation system, pest-repellent plants, traps for insects, and adequate use of the entire periphery. The unit must be aesthetically pleasing and clean, with implements and supplies kept in good order. The points of sale must be clean and hygienic, with produce prices clearly indicated. Items sold by the bunch must weigh at least 1 lb, with the exception of aromatic plants.

GNAU also has the authority to accord municipalities the status of candidacy for referencia nacional status, either for their urban agriculture efforts as a whole or for their success in a specific subprogram or activity, based on criteria listed in the *Lineamientos*. The transition from candidacy to full status of referencia nacional requires the approval of the minister of agriculture. The aspiring municipality must submit to a thorough inspection and validation of results by GNAU. One can get a sense of how demanding and rigorous the selection process for national candidacy is by considering the following 4 criteria, picked more or less at random from the 22 listed in the 2007 *Lineamientos* (GNAU 2007a: 99):

- There must be the highest possible level of uniformity across the popular councils of the municipality in all 28 subprograms of urban agriculture.
- The municipality must strengthen the work of patios and the neighborhood CDRs, thereby ensuring the existence of at least 50 fruit tree species in the municipality (of which 21 are specified in the *Lineamientos*).
- At least 10 varieties of vegetables must be available at the points of sale during all 12 months of the year.
- The municipality must have an apiary producing hives of stingless bees, and must distribute the hives to organopónicos, huertas intensivas, patios, and the municipal seed farm.

In 2007 only 8 municipalities (out of 169) had earned referencia nacional status, and of these, only the municipality of Yaguajay in Sancti Spíritus province had maintained that status continuously since the year 2000. In addition there were four candidate municipalities and a handful of others that had achieved referencia nacional status in individual subprograms such as organopónicos, huertas intensivas, and marketing (GNAU 2007a: 92).

For base-level production units, there are two more levels of promotion. After a unit has been de referencia nacional for two years, GNAU may choose, after what it calls a "profound analysis," to declare the unit *de excelencia* (excellent), and as of October 2007, units with long-standing excellence may be awarded *de excelencia doble corona* (double crown). De excelencia designation requires that, in addition to GNAU, the local government authorities and the state organ under which the unit operates must concur with the designation. Of the hundreds of thousands of base units that existed in the urban agriculture system in 2007—around half a million patios alone were registered—only 94 were designated as units of excellence. These units were scattered across all 14 provinces and Isla de la Juventud. About half were organopónicos and patios. The rest were distributed across 22 other categories, including a day-care center, a vermiculture unit, bee centers, and several UBPCs. Here, as an example of why excelencia status is so difficult to achieve, are the minimum requisites GNAU (2007a: 99) has established for a patio of excellence:

- The patio must have at least 15 of the subprograms present, all of which have to receive a "good" evaluation. By way of exception, however, a particularly outstanding example with even a single activity may be considered.
- The 15 subprograms present must include fruit trees, vegetables and fresh condiments, rabbit raising, chicken raising, vermiculture, composting, and ornamental plants and flowers.
- There must be at least five "fine" species of flowers, such as gladioli, dahlias, petunias, birds of paradise, roses, chrysanthemums, spikenards, and carnations.
- At least five different species of fruit trees must be present, including key lime and dwarf guava.
- Banana trees, if present, must be well kept, according to the accepted practices in banana cultivation.

- The patio must be aesthetically pleasing, with all space utilized and a nice garden in the front or the side. In general it should not give rise to the slightest doubt that it is a patio of excellence.

The last point clearly brings in a subjective element: essentially, the GNAU inspectors need to be convinced that the patio is excellent "any way you look at it."

One more aspect of the political environment in Cuba can also be considered as a direct moral incentive for the urban agriculture movement: the strong commitment to it at the highest levels of government. As noted in chapter 1, then minister of defense and current president of Cuba Raúl Castro Ruz was instrumental in the consolidation of organopónico technology, beginning with its introduction in MINFAR facilities in 1987. His visit to HORTIFAR on December 27, 1997, on the tenth anniversary of the first organopónico, was an important milestone in the urban agriculture movement. He has met with leading urban agriculture producers every year since then on December 27 at the yearly plenary session of the National Urban Agriculture Movement, and has made occasional visits to successful urban agricultural units. Adolfo Rodríguez acknowledges the significance of this support: "The national plenaries taking place with the participation of the leader of the revolution, and the meeting on each December 27 of the best producers of each territory with Compañero Raúl, have been a huge encouragement for the movement" (Rodríguez Nodals 2006: 26–27; my translation).

The collection of material and moral incentives that surround Cuban urban agriculture certainly contribute to a propitious environment for its further development. Whatever other challenges urban agriculture may face in Cuba, now and in the future, in terms of physical or human resource constraints or changes in the economic climate, for the time being the movement is able to take advantage of these propitious circumstances to foster its growth.

6

◇◇◇◇◇◇◇◇◇◇◇◇◇◇

Technological Innovation
in Urban Agriculture

Examples from Protected and Semiprotected Cultivation

The fourth main factor that has enabled the success of Cuban urban agriculture is the country's formidable capacity to adapt and innovate. This theme was woven through chapters 3 and 4, but this chapter illustrates the role technological innovation played in the initiation of the Organoponía Semiprotegida (Semiprotected Organoponic Cultivation) subprogram, which evolved from *cultivo protegido* (greenhouse-based cultivation). In addition to chronicling the introduction and evolution of this subprogram, one important facet in its practice—namely, intercropping—is analyzed.

This recently established subprogram is particularly pertinent to focus on because the way Cuban leaders have introduced it illustrates all the strengths of Cuban urban agriculture in crystallized form. Cubans had long been aware that the quantity, quality, and variety of available produce declined during the hot summer months. Greenhouse cultivation was impractical for feeding the population at large, for reasons described below. The centralized Cuban science (research) and development (technology) institutions, in this case under the leadership of INIFAT, were able to come up with a solution that satisfied both binding constraints on urban agriculture in Cuba: very sparing use of petrochemicals and of imported inputs. The central government also rapidly allocated the necessary investment funds, especially in hard currency, to enable hundreds of hectares of *casas de cultivo semiprotegido* to be spatially distributed across all the island's provinces. Concrete decisions about the actual production process in each location—how to produce the inputs, what seeds to utilize, and what crops to plant—are largely left in the hands of the local growers, who are best able to adapt to local climatic and market demand conditions. Moral and material incentive schemes are in place to motivate intense dedication to the task of rapidly deploying a new technology. The combination of firm guidelines and flexibility that this subprogram exhibits will serve Cuba well in similar efforts in the future.

Precursors in Traditional Greenhouse Cultivation

At the start of the twenty-first century, greenhouse (or hothouse) cultivation of vegetables and fruits is a well-established practice in worldwide agriculture. It has the obvious advantage of local climate control, in terms of temperatures and humidity, to extend or even make possible a growing season. It also offers many other benefits: increased yield; better and more uniform quality of produce; efficient use of water for irrigation; stable, year-around employment for agricultural workers; and the conversion of previously unused or unusable space into cultivable space. Although not an intrinsically urban practice, greenhouse cultivation is well suited to urban settings, where land available for cultivation is scarce and often of poor quality.

This modality also has some serious limitations, however. Because hothouses become attractive and hospitable targets for plant pests, especially soil-based ones, much heavier applications of (typically petrochemical) pesticides are required compared with conventional open-air agriculture. Moreover, to obtain the high yields, it is necessary to use a lot more fertilizer. Extensive use of petrochemical fertilizers and pesticides in an enclosed space in the midst of a population center obviously poses serious potential threats to hothouse workers, urban neighbors, and eventual consumers of the produce. Less direct but just as serious is the potential for damage to the environment, including the aquifer, particularly when hothouse cultivation is extended into fragile ecosystems (Hernández Díaz et al. 2006).

The Cuba of the Special Period faced additional complications for greenhouse-based production. The greenhouses were expensive to build and required many imported components, while cultivation in them depended on heavy use of petrochemical pesticides and fertilizers. Cuba produces little petroleum and has very limited capacity to manufacture petrochemicals. Thus, both greenhouse structures and petrochemicals would have to be imported at costs exceeding world prices due to the U.S. commercial blockade, and they would have to be paid for in hard currency, which Cuba lacked. An additional expected cost was imposed by the high frequency of devastating hurricanes on the island. With hothouse cultivation, not only the current crop but the entire fixed investment in the structures could be lost if a hurricane took an unfortunate course.

Despite all of these drawbacks, cultivo protegido was introduced into Cuba in the early 1990s, mostly but not wholly in urban locations. Why? Not, as might be expected, because Cuba needed to feed its people, but instead, seemingly paradoxically, to earn hard currency for the country. By 1993, the Cuban government had legalized the free circulation of U.S. dollars on the island and had established the TRDs (dollar stores), where all items were priced in U.S. dollars (later Cuban convertible pesos) at levels that created a heavy hidden tax on dollar holders. It had started having success in transforming Cuba into an international tourism destination, despite the U.S. ban on travel to the island. The mostly European and Canadian visitors would expect modern hotels and amenities (which had to be built with Cuban and European capital), and edibles of high and standard quality. Initially, foods such as lettuce and tomatoes had to be imported at considerable cost, reducing the net dollar income from foreign tourists and to some extent defeating the purpose—namely, hard-currency earnings—of introducing tourism as a substantial sector of the island's economy. So, the initiation of cultivo protegido can best be understood as an import-substitution strategy, aimed at increasing Cuban value-added and foreign-exchange earnings. In fact, to this day, the overwhelming share of the output of Cuban greenhouses goes to earning dollars through exportation, provisioning of tourist hotels, and sales in the dollar stores.

Even in this early phase of importing hothouse technology and inputs to substitute for imports of goods consumed in the tourism sector, Cuba's research and development system immediately started seeking ways to increase the value of this effort. First of all, European greenhouse cultivation practices could not be adopted wholesale. The primary concern in Europe is to keep plants warm during the harsh winters (thus the name "hothouse")—not a problem in the tropics. In fact, especially in summer, greenhouses face the opposite problem: the temperatures inside rise high enough to harm both plants and workers (up to 45°C, or 113°F). An enclosed structure is, however, still needed during the tropical summer for humidity control and for fighting airborne viral and fungal diseases (Revista.mes.edu.cu 2007). In 1998, researchers at the IIHLD, working with the Cuban-Spanish joint venture Carisombra, introduced a greenhouse design better suited for the tropics; by 2006 these Carisol greenhouses were being marketed in Mexico, the Dominican Republic, Venezuela, and other Caribbean islands (Pérez Sáez 2006).

The other focus of attention for Cuban researchers and practitioners was the search for locally produced biological plant-pest-control agents to substitute for imported pesticides. One such biological control, already described in chapter 3, is the CIGB-developed HeberNem. By 2007 it had already undergone 15 years of development and testing (including eco-toxicological testing) and been proven effective in combating low to moderate nematode infestation. Between 2004 and 2007 its application in greenhouses increased markedly, from 29 ha to 178 ha (Heber Biotec 2007). In the meantime, researchers at CENSA developed a fungus (Klami C) that, unlike HeberNem, establishes itself permanently in the soil, colonizes plants, and attacks the eggs of nematodes (Revista.mes.edu.cu 2007). Because HeberNem and Klami C attack the nematode at different stages of development, the hope is that the two products can be used in combination, perhaps in a single application.

Unfortunately, considerable hard-currency costs remain. HeberNem and Klami C combat only light or moderate nematode infestations, and imported petrochemical pesticides are required to treat heavy infestations. And although Cuba has almost entirely moved away from methyl bromide usage, its recent favorite substitute (endorsed by the United Nations Environment Programme) is the imported non-ozone-destroying chemical Agrocelone, developed by a Spanish company. Still, the combination of biological pest controls with Agrocelone does represent a considerable dollar savings over methyl bromide alone, as documented at the Empresa Cítricos Ceiba. This empresa has 12.4 ha of greenhouses producing tomatoes, cucumbers, and peppers. In 2003, 80% of the greenhouses were using methyl bromide. By 2006, methyl bromide use had been completely eliminated, replaced by HeberNem and Agrocelone, depending on level of nematode infestation. In 2007 the use of HeberNem as a soil disinfectant in the 23 Carisombra greenhouses saved the enterprise about 20,000 CUCs, reducing the hard-currency costs for Agrocelone from 43,500 to 23,500 CUCs (Revista.mes.edu.cu 2007).

Another dollar-saving and ecologically friendly conversion took place at the Organopónico Vivero de Alamar in East Havana, which has 940 m of greenhouses. In 2006 this UBPC switched to a multipronged strategy to replace chemical inputs in its greenhouses. It used a carefully constructed substratum made up of compost, worm humus, and rice husks in the cultivation beds, along with locally produced or CREE-provided biological pest repellents and pesticides. In addition, it employed lettuce as an auxiliary crop planted alongside its main crops—tomatoes, cucumbers, and green peppers—to act as a nematode-attracting trap-plant. In 2006 and 2007, without using any chemical pesticides or fertilizers, this organopónico harvested total produce yields slightly above those achieved in the pre-conversion years of 2003–2005. Cucumber and tomato yields were somewhat lower, and pepper yields somewhat higher, but with the addition of 2.5 kg/m/year of lettuce, total yield increased from 23 to 23.45 kg/m/year (Navarro 2007).

In sum, prospects seem good for future Cuban greenhouse production to move steadily toward ecologically friendly cultivation techniques that do not require expensive imports. Already, *casas de cultivo protegido* have established themselves as a valuable part of Cuban agricultural efforts. Scattered across both urban and rural Cuba, they play a niche role in provisioning tourist centers and TRDs, thereby contributing to Cuba's hard-currency earnings. But they are hardly a major contributor of vegetables to the Cuban population at large. Production costs also remain high, because even the Cuban-designed Carisol greenhouses require many imported components for their construction. Furthermore, although advances have been made in biological pest control, imported pesticides are still required, entailing considerable hard-currency expenditures, as well as being ecologically undesirable.

Semiprotected Organoponic Cultivation

Within the urban agriculture movement, the focus has shifted in recent years away from cultivo protegido to *cultivo semi-protegido,* away from a modality that basically services the tourism sector and dollar stores, to one that produces vegetables for consumption by average Cubans. This shift is reflected in the fact that the Cultivo Protegido subprogram in the 2005–2007 edition of the GNAU *Lineamientos* has been dropped and replaced by Organoponía Semiprotegida (Semiprotected Organoponic Cultivation) in the 2008–2010 edition (GNAU 2007a).

The new technology referred to as *organoponía semiprotegida* was developed by a multidisciplinary research team at INIFAT. It combines the organoponic, agroecological technology already in place in urban agriculture with plastic agrotextile shade cloths that protect plants from the intense radiation of the tropical summer sun. This effectively extends the season for high-yield vegetable production year-round. In addition, these screens provide at least partial protection against airborne insect pests and damage from heavy downpours. The 2008–2010 *Manual Técnico de Organopónicos* lists several advantages of this technology (GNAU 2007b: 71):

- The hard-currency investment costs are only one-seventh of those required for greenhouses.
- The technology is sustainable, requiring little or no use of petrochemical fertilizers or pesticides.
- Placing these screen covers on a regular organopónico unit greatly enhances protection against heavy rains and flooding.
- In the event of a hurricane warning, it is very simple to dismantle the screen, store it in a safe place, and remount it once the threat is over.
- Because it cuts out 30–35% of the sun's rays, the screening enables much more efficient production during the summer season.

The main novelty introduced in this technique, as contrasted with regular organoponía, are the screens (see fig. 6.1). INIFAT details the recommended installations in the 2008–2010 *Manual Técnico* (GNAU 2007b: 73–75). The smallest recommended size for a semiprotected organoponic unit is 640 m of cultivable area and a gross area of 854 m (including pathways between beds). The largest recommended unit under one roof is about half a hectare (a gross area of 5,184 m and a net cultivable area of 4,608 m). Since the main objective of the screen is to reduce the intensity of the sun's rays, it must be made of a dark transparent plastic agrotextile, tough enough to withstand winds up to 100 km/hour, with reinforced grommets at any attachment points to prevent tearing in the wind. A minimum density of 80 g/m is recommended. The support posts are made of galvanized steel sturdily anchored in cement anchors up to 1.6

Figure 6.1. Semiprotegida organoponía cultivation at UBPC Alamar in Havana. (Photo by author.)

m deep. Galvanized steel tensors connect the posts together, providing a sturdy structure for supporting the screens. The posts are of two different heights—2.5 m and 3.5 m—so that the screens slope to facilitate runoff of rain and optimal sun shading. The agrotextile and galvanized steel structures should have an expected lifespan of at least 5 years, and depending on the manufacturer, the galvanized steel may have a guaranteed life of up to 20 years, even in the tropical climate of Cuba.

Successful deployment of this technology depends on strict adherence to specified agronomic practices, also detailed in the *Manual Técnico* (GNAU 2007b: 71–72):

- High-quality substrate in the beds, with nematode-free soil, at least 50% organic material, and worm humus produced at a nearby location.
- At least six rotations of crops per hectare per year.
- Intercropping in at least 50% of the beds.
- A high level of biodiversity, with at least 10 crops per hectare, not only ensuring a good variety of produce but also deterring pest attacks.

- Four layers of repellent/attractant and barrier plants to deter invading insects around the periphery of the installation and at the ends of each bed, according to the following pattern: a curtain of neem trees 6–12 m from the edges of the installation; a two-layer barrier at the edges of the installation, with *flor de jamaica* (*Hibiscus subdariffe*) on the outside and sorghum or maize on the inside; and repellent plants (at least 50% *flor de muerto* [marigold, *Tagetes evita*] along with purple basil and oregano) at the ends of each bed. In addition, color traps, with appropriate sticky substances, are to be installed on the posts of the installation (50% yellow, 25% blue, 25% white).
- An adequate disinfection station so that pests are not carried in on workers' or visitors' shoes.
- Avoidance of un- or underutilized beds. The motto is "100% population." Plants that die must immediately be replaced either by another transplant of the same species or by a plant with a shorter growing cycle. After each harvest, the bed (full or partial) must be readied and replanted within 48 hours.
- Hurricane preparedness. Prior to a hurricane, the screen and posts must be dismantled and stored to avoid damage. After the hurricane has passed, the screen is to be remounted, the damaged crops removed, and the beds immediately replanted, making sure to include some short-growth-cycle plants such as radishes, lettuce, and Chinese chard so as to ensure new supplies of vegetables as soon as possible.
- A properly functioning system of irrigation. Blockages in the microjets are particularly to be avoided.
- Maintenance of the beds. The passages between them and the periphery of the entire installation are to be free of weeds.
- Proper mix, rotation, and intercropping of crops. Certain crops are to be avoided: Chinese green beans do well during the summer under the open sky; fungus- and virus-prone hybrids and varieties of tomatoes, peppers, and cucumbers do not perform adequately under semiprotected conditions and require greenhouses for their cultivation. Crops such as squash, melon, or chayote would imply an underutilization of this technology and are best grown by conventional agriculture.

GNAU (2007b: 77–78) also offers recommended crop rotations. For a 1 ha plot, the following four patterns are recommended as performing well:

- beets—lettuce—carrots—lettuce—cabbage—Spanish chard
- carrots—lettuce—beets—lettuce—broccoli—lettuce
- broccoli—carrots—Chinese cabbage—lettuce—beets
- cauliflower—eggplant—beets—Chinese chard

For a 1 ha plot divided into four 0.25 ha blocks, the following pattern is suggested:

- cabbage—carrots—lettuce—beets—lettuce—lettuce
- lettuce—cabbage—carrots—beets—Chinese cabbage
- carrots—cabbage—beets—lettuce—lettuce

As with all other subprograms, GNAU evaluates and assesses the performance of organoponía semiprotegida units. For this purpose it has published a set of 13 weighted evaluation criteria, assigned point values that total 100 points (see table 6.1). The productive units are graded as follows: 97 and above = *muy bien* (very good); 90–97 = *bien* (good); 80–89 = *regular* (average); and below 80 = *mala* (poor).

Each of the 13 criteria that contributes to the final grade has a different story to be told about its technological development, adaptation, and incorporation into the process of cultivation. To treat them all comprehensively in this chapter would be impractical, so just one, intercropping, serves as an illustration of the process by which the different strands of the fabric of Cuban urban agricultural programs were spun.

Polyculture, or intercropping, refers to the growing of multiple crops in association, in the same space and at the same time. Historically, many traditional peasant communities worldwide endogenously established this practice, and for good reason. The intercropping of appropriately chosen associated varieties has several benefits. It increases overall yields in both physical and economic terms. Biodiversity makes the plot less susceptible to pest infestations than monocultures are. Associated cultivars can serve as pest repellents or traps, or they can attract natural predators of pests of the main crop. The presence of different root systems with varied characteristics contributes to soil stabilization and conservation and helps retain soil fertility.

Table 6.1. Evaluation criteria for semiprotected organoponía

Indicator	Value (points)	Observations
All beds sown	15	Only beds harvested within the last 48 hours are allowed to be empty
Quality of the substrate	10	Evaluated by observation and examination of its physical properties
Intercropping	5	A minimum of 50% intercropping, with radish, Chinese chard, and lettuce as short-growth-cycle vegetables
Control of pests and disease	10	Evaluated by observation, including of traps and biodiversity
Barriers and repellent plants	10	Visual inspection
Training	5	Exchange with producers
Tying (*vinculación*) of workers' incomes to results	10	Checking documents and exchange with producers
Marketing	5	No produce lost; all go to correct destinations
Irrigation and drainage	5	No blocked sprayers, filters clean, and drainage functioning
Timing of harvests	10	At least five harvests in progress during visit
Population	5	Practical verification
Vermiculture	5	Production is taking place in a nearby area: shade, humidity, substrate
Integralidad, aesthetics, facilities for women	5	Observation
Total	**100**	

Source: Rodríguez Nodals 2007.

Agronomists have developed quantitative concepts to measure the beneficial impact of intercropping. These include physical measures such as per-hectare outputs of total food energy (kcal/ha) or protein (kg/ha), and economic measures such as profits per hectare ($/ha). In terms of the efficient use of land area, the most relevant concept is land-equivalent ratio (LER), which is defined as follows: Each crop has a fixed land productivity ratio (measured, for example, in tons per hectare) in monoculture cultivation. To evaluate the efficiency of an intercropped plot, the yield of each crop in the plot is calculated. The LER is then the ratio of the area that would be required to produce that combination of crops under monoculture to the area that actually produced that quantity of said crops under intercropping.[1] The advantage of LER is that it is a pure number independent of the units chosen to measure area and output. If the LER is greater than 1, more area is required to produce the identical output in monocultures than in intercropping, and the latter is the more efficient modality. If, for example, LER = 1.65, that means 65% more land would be needed to grow the same crops in the same quantities in monocultures than in association.

Large LER values, empirically observed by peasants, go a long way toward explaining the historically and geographically widespread acceptance of intercropping, including in Latin America and Cuba. One study published in the 1980s found that substantial majorities of bean and maize crops and more than 40% of the cassava crop in Latin America were being raised using intercropping (Casanova et al. 2002). In Cuba, intercropping has a long history. In the nineteenth century, sugar plantations had self-provisioning plots for slaves and other farmworkers where intercropping was widely practiced. Outside of sugar plantations, intercropping was used on diversified farms and, in a different form, in fincas growing perennials such as coffee intercropped with shade trees. In banana and plantain production, shorter-life-cycle plants were raised in the spaces between the trees, especially roots and tubers such as cassava, sweet potato, and yucca. The last could also be intercropped with vegetables, which typically have much shorter life cycles, as successful associated crop pairs in general tend to have life cycles of different lengths.

Twentieth-century Cuba saw two developments that diminished intercropping practices considerably. In the first half of the century, the single-minded pursuit of profits on (mostly U.S.-owned) sugar plantations led to the elimination of self-provisioning plots for workers (because slavery had been abolished in 1881, workers were expected to be self-sufficient). Then, in the second half of the century, after the 1959 revolution, Cuban agriculture was transformed into the Soviet model of industrial agriculture, which emphasized large fields with monocrop cultivation. So, for about a century, for different reasons, intercropping receded in importance. It never disappeared completely, however, surviving mostly in perennial-annual or root/tuber-vegetable combinations.

The onset of the Special Period and the concurrent collapse of conventional industrial agriculture refocused Cuban agronomists and scientists on the benefits of intercropping as practiced in the preindustrial, agroecological era. This refocusing meant both a rediscovery process of trying to rescue locally applicable, empirical knowledge based on the historical experience of the Cuban campesinos, as well as a serious research-and-development effort to discover new possibilities for intercropping. The latter path was especially relevant for urban agriculture and its mainstay organopónicos because the primary surviving intercropping practices of the twentieth century in Cuba, namely, perennial-annual and root/tuber-vegetable, were not applicable in organopónicos. Raised-bed cultivation required that suitable associations be found among relatively short-lived vegetables.

The most recent *Manual Técnico de Organopónicos y Huertas Intensivas* (GNAU 2007b) presents the lessons learned from 15 years of intercropping in organopónicos, compiled by GNAU with contributions from various research institutes and input from discussions among grassroots producers, technicians, and researchers. The manual lists technical preconditions for a successful association of crops, including the following (GNAU 2007b: 65):

- The production unit's full program of intercropping for the entire year must be planned ahead of time.
- The principal crops in the associations must be identified.
- All agro-technical work in support of the secondary crop must be subordinated to the demands of the principal crop.
- The associated crop should differ from the principal crop in size and stature.
- Associated plants should be of different biological families, so that they are not susceptible to the same set of pests and diseases.
- The life cycle of the associated plant should be shorter than (or at least different from) that of the principal crop.
- Water and irrigation demands of the principal and associated crops must be compatible.
- Systems of simultaneous planting for the principal and associated crops must be established, depending on the reason and goal for the association.
- Both the principal and associated crops should be planted in

uniform rows, so that the foliage is distributed evenly over the length of the bed.

GNAU (2007b: 66) reports benefits from associations within organopónicos, including increased stability and productivity of the agro-system, better conservation and fertility of the substrate in the beds, reduction in undesirable plants and weeds, and increased profits based on higher yields. These benefits have been quantified. For example, a 1996 study by Caraza et al. (cited in Casanova et al. 2002: 149) found LER values of 1.44 for cucumber in association with lettuce, 1.93 for cucumber in association with radish, and 1.86 for green beans in association with radish.

GNAU has an elaborate list of effective and ineffective associations among vegetables in organopónico beds (see table 6.2). Various authors report other common associations in organopónicos, including lettuce as principal crop associated with radish, chard, leeks, garlic, and onions; cabbage associated with lettuce, chard, garlic, and onions; green pepper associated with radish, garlic, onion, lettuce, chard, leeks, and green beans; bush beans associated with lettuce, chard, garlic, onion, and cilantro; and tomato with lettuce, radish, and cilantro (Casanova et al. 2002; Rodríguez González 2008).

As this chapter has demonstrated, GNAU exercises considerable control over organoponía semiprotegida through its objectives, evaluation criteria, and technical manual. At times this guidance can be quite prescriptive. Numerous practices that have been found unworkable, unprofitable, wasteful, or harmful during INIFAT's research and development of organoponía semiprotegida are prohibited. As an example, the *Lineamientos* prohibit growing Chinese green beans, tomatoes, pepper hybrids, parthenocarpic cucumbers, chayote, squash, watermelon, sugar melon, and Chinese squash in organoponía semiprotegida (GNAU 2007a). Any unit that wishes to experiment with raising any of these crops must obtain from GNAU headquarters explicit, written authorization for a fixed period of time. If any unauthorized crops are detected during the quarterly recorrido, 10 points are deducted from the unit's grade. Yet GNAU does leave local producers considerable discretion in deciding what crops to grow and the preferred size of the semiprotected unit, in light of local market conditions and demand, available terrain, and other local considerations. Here is another example of the combination of strict central supervision with flexible local decision making.

Table 6.2. Possible crop associations in organopónicos and huertas intensivas

Principal crop	Associated crops	Antagonistic crops
tomato	onion, parsley, carrot, lettuce, radish, chard, green onion	cabbage
cucumber	lettuce, radish, onion, beans	none
beans	carrot, cucumber, cabbage, most vegetables	garlic and onion
garlic and onion	beets, lettuce, tomato	beans, peas
broccoli, cauliflower, cabbage	celery, onion, beets, aromatic plants	tomato, beans
peas	carrot, celeriac, radish, cucumber, most vegetables	garlic, onion
spinach	lettuce	none
radish	peas, lettuce, carrot, tomato, green beans, cucumber, pepper	none
carrot	lettuce, radish, peas, tomato, onion	none

Source: GNAU 2007b: table 6.

Dispersion of Organoponía Semiprotegida

Within only a handful of years, organoponía semiprotegida technology has already been implemented in widely dispersed locations across the island, thanks to INIFAT's leadership in generating the technology, pushing for its application, and directly supervising the producers and evaluating the results. An important milestone in the introduction of this new technology was the Seminario Nacional de Cultivos Semiprotegidos, held in February 2007 in Pinar del Río, and attended by María del Carmen Pérez, acting minister of agriculture. At this meeting it was announced that organoponía semiprotegida would henceforth be a subprogram of urban agriculture and would be a high priority for GNAU, with the aim of resolving the problem of the unavailability of many vegetables during the summer months.

Table 6.3. Plan for vegetables and fresh condiments in organoponía semiprotegida, 2008–2010 (in metric tons)

Territory	2008 (t)	2009 (t)	2010 (t)
Pinar del Río	1,575	3,675	4,900
La Habana	1,500	4,500	6,000
Ciudad de la Habana	2,000	4,500	6,000
Matanzas	375	3,525	4,700
Villa Clara	200	3675	4,900
Cienfuegos	825	2,625	3,500
Sancti Spíritus	200	1,425	1,900
Ciego de Ávila	200	1,425	1,900
Camagüey	825	2,475	3,300
Las Tunas	200	1,275	1,700
Holguín	200	1,875	2,500
Granma	825	2,175	2,900
Santiago de Cuba	825	2,175	2,900
Guantánamo	825	2,175	2,900
Isla de la Juventud	600	700	800
Total	**11,175**	**38,200**	**50,800**

Source: GNAU 2007a: 42.

Starting in March 2007, Cuba began investing 2 million CUCs (U.S. $2.2 million) in the construction of 150 ha of organoponía semiprotegida throughout the island, with 30 ha in Havana and 50 under the auspices of MINAZ on land retired from sugarcane production (Suárez Ramos 2007a). Organoponía semiprotegida is a work in progress, and it would be premature to make definitive statements about its results, but examples of successful implementation already abound. By mid-2008 the first MINAZ semiprotegida unit was already in operation in Güines in La Habana province, located on the site of the idled Amistad con los Pueblos sugar refinery. A small group of workers constructed 90 walled beds and an efficient irrigation system. The unit currently has six workers, all of whom are paid by the pago por resultados system. Despite selling their produce at affordable prices, they have managed to turn a profit supplying markets in Havana, a few stands in the small town of Güines, and a permanent stand in the community of Amistad con los Pueblos (Fernández 2008).

Another success is La Quinta, situated near the provincial capital of Sancti Spíritus. Employing 15 highly qualified workers, by May 2008 this unit had produced more than 100 t of vegetables for the tourism sector, social consumption, and sale in the state MAEs. Workers are paid a base wage plus bonus payments for meeting production goals and for correct use of the technology. The technological discipline at this unit is evident in the use of intercropping, worm humus, efficient irrigation and drainage, plant barriers for pest control, insect traps, and biological pesticides. La Quinta has received a perfect grade of 100 in all GNAU recorridos and has earned referencia nacional status two years running. It aspires to reach excelencia status once more varieties of vegetables are introduced and the aesthetics of the unit are upgraded (Jiménez Díaz 2008).

In Havana, the organopónico Vivero de Alamar was allocated 1 ha of organoponía semiprotegida (of the 30 ha planned for the capital). I visited several times during 2007 and observed some functioning units already in production. These examples could be multiplied many times across all the provinces and Isla de la Juventud. During GNAU's thirty-seventh quarterly recorrido in 2007, 27 organoponía semiprotegida units were visited, located in 12 of the 14 Cuban provinces and in Isla de la Juventud. All but three units received grades of good or very good. The remaining three were judged to be average, and not one received a grade below 70%. The 2008–2010 *Lineamientos* gives quite ambitious goals for expanded production, reflecting GNAU's commitment to this technology. As table 6.3 shows, a 242% increase was planned for 2008–2009, and 133% for 2009–2010. In conclusion, the Organoponía Semiprotegida subprogram is up and running, and progressing quite satisfactorily and rapidly.

7

◇◇◇◇◇◇◇◇◇◇◇◇◇◇◇

Case Studies of Urban Agriculture

The constitutive factors of urban agriculture considered in the preceding five chapters express themselves in various ways in the base units where the actual production takes place. From January to May 2007, during a sabbatical semester in Cuba, I visited many such sites. In this chapter I assess the experiences of the grassroots producers based on my personal observations and conversations with participants in urban agriculture.

The majority of my visits were official visits in Ciudad de la Habana, Pinar del Río, and Matanzas provinces arranged with the cooperation of the Facultad Latinoamericana de Ciencias Sociales (FLACSO, Latin American College of Social Sciences), INIFAT, and other organizations, and ultimately with the implicit approval of the Ministry of Agriculture. I remain extremely grateful for the time and attention that Cuban officials, academics, and urban agricultural producers devoted to preparing for and hosting my visits. In addition, I attempted informal and impromptu visits to production units in Havana and in Santiago de Cuba during a brief stay in the latter locality. Most of these informal visits consisted of "drive-by" or "walk-by" visual observations. In some cases I attempted to enter the units and contact the people in charge. I was politely refused access for two understandable and good reasons. For one thing, no agricultural-ists would appreciate having their daily routine disrupted by the unan-nounced visit of a foreign academic. The other reason has a more sinister history. Cubans believe that the United States has subjected their country to biological warfare over the last 50 years in an attempt to disrupt and damage its food production systems. It stands to reason, then, that they would not open the doors of their production facilities to an unknown, visiting U.S. scholar. Even though I was not allowed access to the fields, the people in charge whom I encountered were generally willing to have informal conversations and answer some questions about their facility or about urban agriculture in Cuba and their roles in it.

A few of the units I formally visited were freestanding patios or parcelas, essentially operated by a single family and perhaps a few associates. Most, however, were examples of cooperative or collective production: UBPCs, CCSs, and production units belonging to state enterprises. Of course, visits to CCSs typically involved tours of not only the collective cultivation area but also of the parcelas of individual CCS members.

Perspectives of Leaders of the Urban Agriculture Movement

Each official visit was usually preceded by a preparatory meeting at the institution—for example, FLACSO, INIFAT, or ACTAF—that arranged the visit. The most comprehensive of these were two lengthy interviews at INIFAT, the headquarters of GNAU, with Dr. Nelso Companioni Concepción, vice-president of INIFAT and executive secretary of GNAU. The consensus among the officials I spoke with during these meetings is that the introduction of urban agriculture into the Cuban economy has entailed the creation of new kinds of economic and social actors who are significantly different from the historical participants in the post-1959 Cuban agricultural sector. After the revolution, the agricultural workforce was divided into two categories: salaried workers in state enterprises, and cooperative (CPA or CCS) members or individual landowners living in rural areas of Cuba. Their activities were strictly controlled by central government policies in a planned economy, the state having monopoly and monopsony powers in relation to private producers or cooperatives. The state assigned binding production programs to producers through contracts at state-fixed prices. The state also provided all inputs needed in production and purchased all commercial output.

Urban agriculture, as it has evolved in the Special Period, has produced a workforce and an economic and social environment with very different characteristics. First, the workforce is recruited in urban areas, mostly from among urbanites who are highly educated and culturally more sophisticated than their rural counterparts. Despite all the Cuban government's efforts over the last 50 years to equalize rural and urban opportunities, cities have continued to offer more and better cultural services (theater, ballet, museums, bookshops) than rural areas do. As Dr. Companioni casually noted in conversation, he knew a couple of university professors in philosophy who now run an organopónico. Of course, being well educated also implies being capable of absorbing education and training in a new field, such as agricultural technology.

Second, in the post-1990 environment of Cuba the central authorities are significantly less capable of running a centrally planned agricultural system. Meanwhile, the highly educated, culturally sophisticated workforce in urban agriculture is a lot more capable of taking initiative and making autonomous decisions. As a result, the directives from the top are a lot less prescriptive than in the past. In general, decentralization has meant that local people have to be persuaded, not ordered. If the central authorities wish to introduce an innovation, they typically do not try to force it on recalcitrant producers, but rather seek to enlist their cooperation through persuasion. According to Dr. Companioni, the local *alcalde* (mayor) is the first person GNAU has to get on its side when introducing an innovation to a new area. The consultorio-tienda system also plays an important role in diffusion. Dr. Companioni gives the following example: the avocado variety currently most common in Cuba, the Catalina, produces fruit only in August and September. INIFAT has developed varieties that have seasons lasting four months, that mature in different seasons, or both, making year-around avocado availability feasible. If a grower comes to the local CTA intending to buy seedlings of the Catalina variety, the extension agent will not dictate what the grower may buy, but will try to persuade him or her to buy a mix of Catalinas and some new variety (or varieties) for experimentation.

Third, almost uniformly, urban agriculturalists operate in an economic environment in which their incomes are tied to their achieved results (*ingresos vinculados con resultados alcanzados*). This economic principle, along with consciousness raising/moral incentives and training, form the tripod that supports urban agriculture. Dr. Companioni calls it the *motor impulsador* (driving engine) of urban agriculture: the producer in the base unit must be motivated, well trained, and well remunerated.

Finally, the proponents and promoters of urban agriculture recognize that is much more than a system for producing food. At least as important as the tons of output achieved are its successes in environmental protection stemming from its agroecological nature, from purposeful introduction of increased biodiversity into cultivation practices, and from enhanced attention to gathering and caring for natural enablers of agriculture: soil, water, and materials for organic fertilizer. Perhaps as important is how urban agriculture has influenced its practitioners and the communities where they live. The leadership of the urban agriculture movement is committed to raising the self-esteem and societal esteem of urban agriculturalists through "human resource" policies, which in Cuba fall under the general heading of *atención al hombre* (attention to man). These policies are designed to empower and "dignify" the men and women who labor in urban agriculture. The urban agriculturalists are educated within and outside of their production units. They are encouraged to continue their formal education and pursue university degrees. They are offered better working conditions and incomes than most urban workers. Their achievements are publicly recognized by GNAU and other government entities, and perhaps more importantly, by the community.

In this context, Dr. Companioni sees urban agriculture advancing in three directions:

1. Endogenous development based on local initiative.
2. Consolidation of agroecological practices that minimize the use of petroleum and its derivatives.
3. Increasingly sustainable practices concerning natural resources: soil conservation, improved capture and utilization of water for irrigation, and better and more complete processing of organic materials for crop fertilization.

The government's role in this development process is, of course, not to be underestimated. This strong central support for urban agriculture has its roots in a 1987 decision by then Minister of Defense Raúl Castro to encourage, despite considerable opposition at the time, the development of its underlying technologies. Today the government remains the main and indispensable source of support for urban agriculture. It distributes public lands in usufruct to willing producers. It supports them with networks that provide seeds, CREE products, and organic fertilizers, in addition to helping base units acquire their own capabilities in these areas. Finally, it educates growers and facilitates the diffusion of innovations through a network of units of *referencia* and *excelencia*. ACTAF, for one, has deployed its membership of more than 20,000 experts on a volunteer basis to promote rapid innovation and diffusion. And, of course, GNAU leaders and their rigorous oversight and supervision play no small part in propelling forward the national movement of urban agriculture. GNAU is keenly aware of its role and stewardship. Hanging on one wall of Dr.

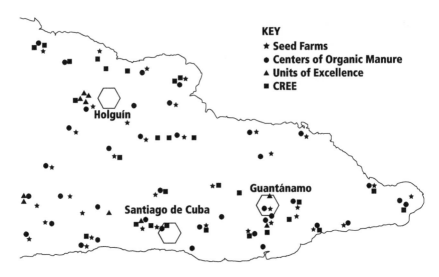

KEY
★ Seed Farms
● Centers of Organic Manure
▲ Units of Excellence
■ CREE

Map of the Eastern End of Cuba

Figure 7.1. The different points on the map represent centers of organic manure production, municipal seed farms, CREEs, and production units "de excelencia." (Drawing by Amanda DeLorenzo, Dickinson College.)

Companioni's office at INIFAT is a huge map of Cuba, with pins of different colors marking the locations of the 600-plus municipal seed farms, municipal organic fertilizer centers, and CREEs. I was allowed to photograph this wall map. It took eight frames to cover the entire island. Figure 7.1, based on one of these frames and showing the extreme eastern end of Cuba, the location of Santiago de Cuba and Guantánamo and the most remote centers from Havana, illustrates how widely dispersed and universal the Cuban urban agriculture effort is, and also how justifiably proud GNAU is of its and the movement's achievements.

Visits to Urban Agriculture Sites

With this background in place, I now turn to descriptions of the various urban agriculture sites I was able to visit while in Cuba. I start with visits in Pinar del Río and Matanzas provinces, officially arranged by GNAU. On both visits I was accompanied by a colleague from FLACSO.

Municipality of San Cristóbal

San Cristóbal is located in the province of Pinar del Río, about 70 km west of Havana. My colleagues and I spent almost an entire day there, visiting various urban agricultural venues in the municipality. We were hosted very graciously by the *jefe de agricultura urbana,* Adolfo Salgado Carrillo, who initially received us at his office at the Empresa de Cultivos Varios de San Cristóbal, where he briefed us about urban agriculture in the area.

The municipality has a surface area of about 1,000 km² and a population of slightly more than 70,000 living in 56 settlements and 12 popular councils. The urban agriculture program boasts 60 organopónicos covering 26 ha. Its one greenhouse was destroyed in a hurricane, and in keeping with the developing trend in Cuba, it has been replaced by a unit of organoponía semiprotegida. In 2006 the municipality as a whole produced more than 34,000 t of output, and the organopónicos alone 580 t.

There is also a hydroponic unit covering 4 ha. Established in 1972, it survived the Special Period transition and was still functioning in 2007 with the tourism sector as its main consumer. Another recent major change has come from the decision to scale down sugarcane production. Although San Cristóbal is still home to an active sugar mill, MINAZ has moved about 100 *caballerías* (ca; 1 ca = 13.1 ha) out of cane production to other uses. Sugarcane workers are being retrained intensively and successfully, including in urban agriculture. Finally, the municipality is a significant egg producer, supplying about 30% of the provincial output and serving population centers as far away as Havana. It is also host to two IPAs, which is unusual among Cuban municipalities. About 30% of the urban agricultural output goes to self-provisioning and social services— schools, hospitals, military units, maternity homes, and Gastronomía Familiar.[1] The remaining 70% is sold to the public, either at the production units or at points of sale scattered throughout the municipality. The urban agriculture program, presumably through the granja urbana, delivers the produce to the points of sale using, among other modes of transport, horse carts. All sales are at prices determined by the administration of the popular council. Even so, none of the UBPCs in the municipality incurred losses in 2006, and monthly per capita incomes for members ranged up to 900 pesos (400+ pesos in salary and the rest in year-end distributions of profits in the cooperative). At the time of our visit in April 2007, the

Figure 7.2. One of the organopónicos at the IPA Batalla del Rubí. (Photo by author.)

municipality had decided to apply to GNAU for referencia nacional status and was working hard to meet the criteria for acceptance, including increased production. In a several-hour-long recorrido following the initial meeting at the empresa, we visited the following three sites, plus the UBPC El Mango.

Hydroponic unit. This 4 ha facility produces mostly tomatoes for the tourism sector. Founded in 1972, it was organized as a UBPC in 1993. Since a reorganization brought in new management in 2005, it has been profitable. At the time of our visit it had 18 workers, all cooperative members. In one sense, it represents a throwback to pre-urban agriculture times. The production here is based on typical hydroponic technology using liquid chemical fertilizers and a nonorganic substrate to support the plants. Since its products earn income in CUCs, it has weaned itself off MINAG subsidies for imported fertilizers and is, in effect, self-sustaining. Still. it cannot be said to be emblematic of the new agroecological urban agriculture movement.

Instituto Politécnico Agropecuario Batalla del Rubí. This IPA enrolls about 600 students between the ages of 15 and 18. Almost all are from the municipality, and those who live nearby go home every night. The rest live on campus and get leaves every 15 days to go home. It has organopónicos that supply its dining halls with fresh produce and that also produce vegetables for social services (fig. 7.2). This school is a net urban agriculture provider: it produces its own worm humus *and* the next generation of urban agriculturalists.

The sugar mill. We drove by but did not visit the functioning sugar mill. Relevant to urban agriculture were the *cachaza* piles, visible from a distance as a row of small hills. Cachaza, the residue left after the filtering process employed in the mill, constitutes an excellent source of organic fertilizer for organopónico and huerta intensiva beds.

UBPC El Mango, San Cristóbal

Because UBPC El Mango is a UBPC de excelencia that has received national recognition at the highest levels, I discuss it separately and in some detail. El Mango has a high-profile public history that both precedes and succeeds my visit to it. The UBPC with this name was founded in 1998 through the initiative of a veterinarian named Nardo Bobadilla, affectionately known as "El Médico." He and 16 others decided to undertake the task of recovering the remains of a failed pork-raising operation called El Batey, in a setting surrounded by 10 ha infested with the noxious weeds aroma and marabu (*Dichrostachys*). The UBPC joined the urban agriculture movement in 2000, and through hard work, determination, and a "sí, se puede" attitude it had already earned de excelencia status by 2003.

El Mango received widespread attention after a visit by Raúl Castro in March 2007, the month before my visit. Castro had high praise for what he observed during his visit, acknowledging Nardo Bobadilla as a magnificent administrator who had "demonstrated with his iron will that in the matter of food production—yes, you can!" He congratulated Bobadilla in the name of Fidel Castro, and promised a return visit (Mayoral 2007).

At the time of my visit, the UBPC—continuously profitable since its inception—had a network of 5 fincas spread over 120 ha, with resident and responsible jefes de finca. Each jefe lives with his family in a house built on the finca by the cooperative and given to him in usufruct for as long as he can keep the finca productive. The UBPC had 64 members, all being paid according to the pago por resultados system. They were growing bananas, plantain, squash, sweet potato, beans, lettuce, beets, radishes, onions, green peppers, chile, and green onions; and raising chickens, pigs, goats, sheep, and rabbits. They sold all their produce at prices established by the Provincial Administrative Council of Pinar del Río. The products were distributed through the state MAEs and also, by contract, to maternity homes, nursing homes, a primary school with semiresident pupils in the nearby community of Modesto Serrano, and Comandante Pinares Hospital in San Cristóbal.

On our visit we were received by UBPC President Nardo Bobadilla and taken on a tour of the facility, with stops at two of the fincas. By this time, only a month after Castro's visit, the co-op membership, all recruited locally, had grown to 76. In 2006 the UBPC had a net profit of more than 400,000 pesos, 60% of which was distributed as income to the members, netting workers an average income, including salary and bonuses, of 800 pesos per month. Construction is ongoing. We saw a huge grader at work constructing a road to connect fincas 4 and 5.

For vegetable production, the UBPC has three micro-organopónicos, to use Bobadilla's term (fig. 7.3), as well as a 0.5 ha organopónico semiprotegido that uses the same sun shades used for tobacco, a prominent crop of Pinar del Río. Water for irrigation is plentiful, as the UBPC has access to two small reservoirs conveniently located nearby. In fact, this was the one advantage the UBPC enjoyed from its inception. Fruit trees line the roads. Central areas of the fincas are dotted with potted ornamental and

Figure 7.3. An organopónico at UBPC El Mango. (Photo by author.)

Figure 7.4. Pigs in El Mango, feeding on seeds from nearby palms. (Photo by author.)

fruit trees, which are later transplanted into the ground to mature. The UBPC has started growing peach trees, a rarity in Cuba. There is a nursery for members' own use and for sales to outsiders.

The UBPC currently raises about 600 pigs under a state contract. For every 20 t of feed the state provides, it receives 5 t of live pigs in return. The UBPC adds locally produced feed supplements and receives sufficient payments from the state for Bobadilla to call it "a good business." The co-op also has 650 chickens, mostly for egg production. At one of the fincas a herd of goats was visible, along with a very clean, well-kept pigsty where additional pigs outside the state contract are being raised in an effort to increase profits. The pigs get part of their diet from the seeds of very tall palm trees that grow nearby (fig. 7.4). The seeds are at the tops of the trees, and the UBPC has to contract a specialized workforce to harvest them. All animal waste fertilizes crops: solid waste for banana trees, and liquid waste for maize. No waste is allowed to go to waste!

Back at his office, Bobadilla printed out from his computer a list of 14 fundamental principles that underlie the operations at UBPC El Mango. They are worth recording here, as they presumably have contributed to the enterprise's success:

1. Care for and tend to members of the cooperative in all senses (atención al hombre).
2. Produce in a strictly cooperative manner, and increase diversification and efficiency every day.
3. Always try to diminish costs.
4. Achieve profitability without ever violating social aims.
5. Never expand administrative staff (*indirectos*), and practice multitasking whenever possible.
6. Establish systematic controls, and critique poorly done jobs at the appropriate time and place.
7. Provide personal and collective rewards for every success achieved.
8. Put one's heart into every assigned task.
9. Make sure that training is part of the preparation of each and every member of the cooperative, especially of area leaders.
10. Ensure that the UBPC functions in all senses—productive, political, social, and cultural—on the basis of unified actions of the administration, political organizations, and people's organizations present in the UBPC, without ever deviating from the mission and assigned role of each one.
11. In order to achieve sustainability, continuity, and systemization in outcomes, . . . insist on organization, discipline, control, training, and the availability of ample, clear, accessible, current, and precise information at each meeting of the workers' assembly.
12. Comply strictly with all general and internal regulations that apply to the UBPC.
13. Make sure that the administrator and the board of directors possess the capacity to think ECONOMICALLY. [Capital letters in original document.]
14. Always propagate the message: Yes, You Can! (My translation)

One example of economic thinking (item 13) was provided by Bobadilla himself in talking about his philosophy of investment: "If you put down five, you should expect to get back six—in exceptional circumstances maybe only five, but never four!"

Since my visit, El Mango has continued to earn laurels. In October 2007, GNAU introduced a new category of excelencia called *doble corona* (double crown). The specified criteria for this designation name 10 activities with required achievement levels in each: an orchard of fruit trees, with a minimum of 30 varieties; an organopónico of at least 0.5 ha or a huerta intensiva of at least 1 ha; rabbit raising with at least 50 breeding rabbits; sheep raising with a minimum herd of 200; an intensive fattening system for at least 50 head of sheep; milking goats, at least 50 head; a herd of goats for fattening; a contract for raising pigs efficiently, with effective disposal of waste; a composting facility with a capacity of no less than 500 m^3; and, finally, a vermiculture center no less than 0.5 ha in size. To earn the doble corona, a unit would have to achieve at least five of these targets by October 10, 2008, according to an established timetable. El Mango had met all 10 targets by the time of the next GNAU inspection during the forty-first recorrido on July 15, 2008. Not resting on its laurels, it had introduced two new activities: the recovery of yellow malanga (a root vegetable) and the raising of quail. Thus El Mango became the first UBPC to be awarded the recognition of excelencia doble corona—and did so ahead of the deadline (Pérez 2008).

The year 2008 also brought its share of adversities, in the form of two hurricanes that devastated Pinar del Río. At El Mango the storms destroyed crops and ripped off the roofs from buildings housing animals, leaving the animals exposed to the elements with little chance of survival. By constructing temporary roofs out of palm fronds that had been brought down by the wind, the UBPC was able to save all of its pigs and most of its chickens and rabbits. Within 40 days of the hurricanes, vegetables were once again being marketed out of the organopónicos. One consequence of this rapid and creative response was that the UBPC closed 2008 with profits of more than 500,000 pesos, with an average monthly income of 850 pesos per capita for its by then 92 members. Two visiting reporters could not resist the temptation of unofficially awarding El Mango the (nonexistent) status of "triple corona" for its efforts to recover from the hurricanes (Suárez Rivas and Suárez Ramos 2009).

Municipality of Matanzas

The municipality of Matanzas (in the province of the same name) lies on the Caribbean coast about 120 km east of Havana. Containing the city of Matanzas and surrounding areas, it covers about 315 km^2 and has a population slightly under 150,000. It was the site of my second official visit to an urban agriculture unit outside Havana.

During this visit we were hosted by Brígido Sergio Alonso Casentes, the agronomist of the municipal granja urbana. Initially trained as a soil agronomist, he has now become a specialist in vegetables and fruits. He was one of the professionals who cofounded GNAU in 1994, and he has been working in urban agriculture ever since. As he pointed out, this kind of professional presence is the norm in Cuba. Each municipal granja urbana in Matanzas province has at least one qualified agronomist such as Alonso Casentes on its staff.

Alonso Casentes escorted my colleagues and me to visit two organopónicos: one at Empresa Forestal, the other Organopónico Triangulo. While driving between the two sites, he filled us in on some aspects of urban agriculture in Matanzas municipality. The municipality possesses 57 organopónicos, more than 50 huertas intensivas, 3,000 parcelas, and 1,800 patios. In the first three months of 2007 the nearly 800 ha of urban agricultural land produced approximately 7,600 t of output, which would correspond to an average annual productivity level of 9.5 kg/ha across all the modalities. (Of course, productivity is somewhat lower in the hot summer months.) In terms of training, 100% of urban agricultural workers receive instruction in relevant technologies. This is an absolute necessity for the system to function. Alonso Casentes himself was shortly planning to attend a three-month ACTAF course for specialists in agronomy.

The organopónicos in the Matanzas municipality, 57 in number and covering 16 ha, grew nearly 930 t of produce in the first three months of 2007. An additional 20 ha of organopónicos were in various stages of construction. In the few kilometers we traveled between site visits, we passed by four organopónicos (of 0.5, 1, 1.5, and 1.2 ha in size) that were at different levels of readiness. One had lost its greenhouse to a hurricane, and only some support posts were left standing.

The effort in Matanzas municipality looks and is impressive, but lest one be misled to believe that urban agriculture in Cuba is concentrated in major cities, only about 23% of the area of urban agricultural land in the province is located in Matanzas municipality. At least one of the 13 other municipalities in the province exceeded the Matanzas municipality in total urban agricultural output in the first three months of 2007.

All production units in Matanzas must grow at least 10 different crops. One result is that Matanzas city is now self-sufficient in all produce except for cabbage, squash, and melon. Prior to the introduction of urban agriculture, all produce had to be brought to the city from fields 70 km away, incurring losses through deterioration and spoilage.

Strict quality and phytosanitary controls are in place. These are reviewed monthly in meetings at the municipal and provincial levels involving producers and consumers, such as MINED, MINSAP, and Acopio (the state wholesale buyer and distributor). Because end users can refuse to accept or pay for low-quality products, problems are identified and resolved rapidly. The produce is sold to the public at the production units. In addition, the granja urbana has a network of 10 points of sale throughout the city and has its own transportation to pick up the produce at the various production sites and deliver it to the points of sale.

Organopónico Triangulo

This organopónico (shown earlier in fig. 2.1) is a subunit of the granja urbana. It is 0.6 ha in size and has four workers tending it. They earn a monthly income of 750 pesos, including salary and incentive pay based on profits. At the time of our visit the unit was growing carrots, sweet corn, okra, cabbage, parsley, and green beans intercropped with lettuce and beets. It had noni trees for pest deterrence. Purple basil and marigolds were being used as repellent plants, and maize as an attractant and barrier plant. The municipal water utility delivered a contracted amount of irrigation water to a cistern on the grounds. The organopónico had plans for vermiculture, but at the time of my visit the beds contained 55% organic fertilizer, mostly cachaza.

The typical yields were 12–15 kg/m^2 annually, although the unit had achieved exceptional harvests of 20–25 kg/m^2 and had grown a head of leaf lettuce weighing 1 kg! About 90% of the output is sold to the public at the unit itself, from a small stand located at one end of the organopónico. The remaining roughly 10% is distributed to social services by contract at preferential prices: a local school for profoundly deaf children (40 lb per day), a day-care center (80 lb per day), and a maternity home (20–25 lb per day).

Organopónico of the Empresa Forestal

This unit belongs to the Empresa Forestal (State Forestry Enterprise). It covers 1 ha and has four workers assigned to it. Situated next to an Empresa Forestal tree nursery, it started as a self-sufficiency plot for the

Figure 7.5. The organopónico in the Empresa Forestal in Matanzas. (Photo by author.)

forestry workers. Now it also contracts with hospitals and military units with the aim of supplying social needs. It makes no direct sales to the public. All of its income is generated through the contracts it has with public institutions to provide produce at modest prices. It grows lettuce (including for seed), green onions, onions, carrots, green beans, Swiss chard, eggplant, beets, and other crops (fig. 7.5). For fertilization it has turned to its own compost pile, and it has plans to introduce vermiculture. The water for irrigation comes from a well. Its raised beds contain about 40% organic material.

This organopónico has achieved annual yields of 18 kg/m^2, and in one bed it raised 8 kg/m^2 of black-seeded lettuce in 35 days. These productivity levels led to a net income of 45,000 pesos in the first three months of 2007, with 14,000 pesos in March alone. The four workers receive a base salary of 500 pesos a month from the Empresa Forestal. In addition, each receives 1,000–2,000 pesos per month as his share of the organopónico's net income, depending on the harvests. According to Casentes, the four workers are allowed to earn as much as their hard work and diligence can generate, with no income cap. No one interferes with their activities as long as they continue meeting their contractual obligations. If they were no longer able to fulfill their contracts, more workers might be added to the unit to increase production, thereby diluting the workers' per capita incomes.

During the visit I could not help but notice billboards brimming over with local pride: "Matanceros—Laboriosos, Inteligentes, Cultos" (Matanzans—Hardworking, Intelligent, Cultured). Judging by their achievements, the eight workers in these two organopónicos surely fit the first two characteristics claimed for Matanzans. It is a shame we could not spend more time with them in the "Athens of Cuba" to explore the third claim.

Santiago de Cuba

In addition to these two formal visits, I tried wherever I went in Cuba to gain and record impressions of local urban agricultural activity. One such opportunity arose when I was able to visit Santiago de Cuba for a couple of days. Santiago de Cuba, the province containing the city of the same name (Cuba's second largest), at the eastern end of the island, has participated fully in the national urban agriculture movement. One of its municipalities, Palma Soriano, has earned referencia nacional status, and the province boasts more patios de excelencia than any other province (GNAU 2007a: 92–96). Some of the city's spacious avenues are lined with organopónicos, huertas intensivas, parcelas, and UBPCs. Along a short stretch of one such avenue, Avenida de las Américas, I observed many functioning urban agriculture sites (fig. 7.6).

Among them is the Organopónico Luis M. Pozo, covering an area larger than 1 ha and announcing on the sign that identifies it that it has the support of the Asociación Cubana de Producción Animal (ACPA, Cuban Association of Animal Production), the European Commission, and the German humanitarian organization Deutsche Welthungerhilfe (DWHH, German World Hunger Aid); and is a part of the Project to Strengthen Urban Agriculture (fig. 7.7). Another UBPC that stood out was the Roberto Alegre Capriles, where we stopped and talked with one of the co-op members briefly. This profitable co-op has more than 50 co-op members and several hydroponic and organoponic units. Santiago de Cuba was the last location where I was able observe urban agriculture in action outside of Havana. I now turn to the sites I visited in the capital.

Figure 7.6. An organopónico along Avenida de las Américas in Santiago. (Photo by author.)

Figure 7.7. The sign identifying Organopónico Luis M. Pozo in Santiago. (Photo by author.)

UBPC Organopónico Vivero Alamar

Located in Havana, this unit provides an outstanding example of urban agricultural excellence. I was first invited to visit it in 2003, when I was in Cuba with 16 U.S. students on an academic stay at the University of Havana. I returned several times in subsequent years. During my sabbatical stay in 2007, I visited it five times, spending long hours interviewing people and observing the daily activities in the co-op.

This unit is located at the extreme western end of the municipality of Habana del Este, in the "bedroom" community of Alamar. It is somewhat mistakenly labeled an organopónico, as it is mostly a set of huertas intensivas and some greenhouses (see fig. 2.2). It was started in 1997 by Miguel Salcines, a mid-level Ministry of Agriculture agronomist. He asked to be given in usufruct a 3.7 ha plot of then unutilized "wasteland," right across the street from a large collection of four- and five-story apartment buildings. He joined forces with four others, including a carpenter and a chemist, to begin founding a production unit. It is fair to say that what has happened since has surpassed all reasonable expectations. From its humble beginnings it has risen to become one of 94 units in all of Cuba to achieve and maintain the classification of excelencia. Between 1997 and 2007 its workforce grew from 5 to 147, its area expanded from 3.7 to 11.2 ha, and its annual production jumped from 20 to 240 t of vegetables.

The land under cultivation at the UBPC was not increased through land purchases, given that there is no land market in Cuba, but rather through two different types of acquisitions. First, an adjoining area of more than 6 ha was absorbed into the production unit as individuals who had been cultivating these lands in usufruct started abandoning their plots because of ongoing improvements in other branches of urban economic activity. This new area is currently being developed into additional huertas intensivas. The second addition came about when the government decided to hand over a state organopónico of 1.2 ha nearby but not adjacent to Vivero Alamar, due to problems in maintaining stable and adequate production levels. The 27 state workers from the old unit were incorporated into the UBPC as cooperative members. They represent only part of the increase in the workforce, as the cooperative had been expanding its workers anyway, in order to increase production and diversify its efforts in cultivation and marketing. New workers are hired for a probationary period of three months, at the end of which they typically become members of the cooperative. Thus, about 135 of the almost 150 people working at the UBPC are members of the cooperative. The fields produce a great variety of vegetables. The main crops are lettuce, Swiss chard, tomatoes, cucumbers, and cabbage, followed by beets and eggplant; ancillary crops are carrots, green beans, celery, cauliflower, mint, parsley, okra, and green peppers. The unit will also soon have a mushroom-growing facility. A coop for more than 200 egg-laying chickens is also planned. Both were under construction at the time of my visit in 2007. When finished, the coop will contribute eggs to sell and chicken manure for fertilizer.

Around 90% of the produce is sold directly to the public at five stands on the street adjacent to the fields at prices set by the UBPC (ranging from 1 to 3 pesos per pound). The UBPC also produces and sells 167 varieties of ornamental plants, as well as fruit and other tree seedlings for transplantation. The remaining 10% or so of the vegetable produce fulfills social needs, being sold to schools and hospitals at reduced prices under contract, or sold to the tourism industry to earn hard currency for the MINAG, which the ministry can then spend for imported agricultural inputs. For example, 90% of the lettuce crop was sold to the public at the retail stands at 3 pesos per pound, 5% was destined for social consumption at 2 pesos per pound, and 5% was purchased by hotels such as Hotel Nacional and Habana Libre at 0.8 CUC/kg.

With this brief snapshot of the unit's work in mind, it is worthwhile to consider how it participates in the four important areas of policies and practices discussed in earlier chapters: training and education, research and development, provision of inputs, and moral and material incentives. In terms of education, the workforce is already highly qualified—albeit not necessarily in urban agriculture! Of the 147 workers, 50 have either engineering degrees or mid-level formal technical training. During their three-month probationary period, new workers go through an intensive training process consisting of on-the-job training, working alongside experienced current members, as well as formal classroom instruction. The unit has a centrally located classroom that also serves as the dining hall for the daily lunches and snacks. On one of the days I visited, a class in English was announced on the whiteboard of the dining hall/classroom without walls. Although this UBPC does not have formal ties with any school circles of interest (described in chap. 4), it maintains an open-door policy, giving consultation and advice to schoolchildren sent there by their teachers to complete assigned projects, as well as to adults in the community with questions about urban agriculture.

The spirit of participatory research and development is alive and well at the unit. Experimentation with and introduction of new technologies take place routinely. During my visit, "liquid smoke"—a bio-preparation made of mango, neem and noni leaves, and fruit—was being tried as a pest control on various crops. As described in chapter 6, intercropping is another focus of urban agricultural technology. In one of the new huertas intensivas on recently acquired lands, an Israeli hybrid tomato was sown experimentally in alternating rows with a cabbage that had previously worked well when intercropped with lettuce and carrots. A nearby "control" bed had the tomato planted alone. This informal experiment revealed that the tomato was growing better alone than intercropped. Among other new technologies that have been or shortly will be introduced are the use of magnetized water for irrigation and the direct rooting of fruit tree grafts.[2] Also scheduled to be introduced in 2007 was 1 ha of semiprotected cultivation (one of the 30 ha allotted to Havana). Because this technology still requires the importation of irrigation systems, its introduction is subject to delays associated with hard-currency availability.

The unit produces some of its own seeds and most of its own seedlings. It purchases from a CREE contracted amounts of biopesticides, which are delivered to the huertas. It produces all of its own compost and worm humus using organic residues of its own operations as well as manure purchased from a nearby farm that raises livestock. With the exception of the portion that until recently was a state organopónico, the unit receives no irrigation water from the city utility pipelines. Instead, it has constructed six wells that provide adequate water for the crops in the unit's original contiguous lands.

A moral incentive for the unit as a whole is provided by its status as a center of excellence, a status awarded to only a handful of units across Cuba. It has maintained this high status through a succession of quarterly GNAU inspections. At the unit itself, the underlying philosophy of atención al hombre ensures that incentives—both moral and material, and individual as well as collective—are sufficient to attract a qualified and stable membership. Dignified working conditions include a seven-hour workday (7 a.m.–3 p.m. with a lunch break); adequate, air-conditioned restroom facilities; and a wholesome lunch consisting mostly of ingredients grown at the unit itself. On one visit I was invited to lunch; the menu consisted of black bean soup, cornmeal with chunks of pork, tomato and lettuce salad, and a cooked tomato-cabbage side dish, with the last two dishes using vegetables grown at the unit. There are also on- and off-site opportunities for members to further their formal education: seven members, for example, were taking classes at the university exten- sion in their municipality at the time of my visit. Members also participate in cultural programs in the local casa de cultura and, as urban residents, can take part in organized sports or enjoy a ballet or theater performance in the evening. The unit leadership is committed to doing what it can to strengthen the perception that urban agricultural work is based on sci- ence and technology in order to support workers' self-esteem and status in society.

Encouraging a sense of ownership among the members is part of treat- ing them with the respect they deserve. Not only are their incomes de- pendent on receipts generated through sales, but the general assembly of the cooperative, consisting of all members, decides on the distribution of this income, following both general laws that apply to all UBPCs and their own internal regulations. The unit's financial affairs are completely trans- parent to the membership. In fact, a blackboard on the wall of the main office displays financial information about the last month's operations, cumulative figures for the year to date, and how well the unit is doing in terms of meeting its annual projections. Displayed are total receipts, to- tal expenditures (including all salaries), cost per peso of receipts, profits, average salary, and average income for the unit (members receive a base salary, and about half of the profits are distributed as individual income; the rest goes to collective expenditures, including investment).

Summarizing the figures on that blackboard during March 2007 yields a sense of the material incentives offered to co-op members. The projected average annual income to be distributed to workers was 8,528 pesos, with 1,421 pesos in earnings for the first two months. The actual average earn- ings for the first two months was 1,629 pesos. Thus, the actual monthly income of the membership, amounting to about 815 pesos, exceeded the projected monthly average of 711 pesos by 15%, and exceeded the average monthly income of salaried state workers, which was 385 pesos, by 112%. It is no wonder that a 2005 study on urban agricultural production units found that 35 of the then 50 members of UBPC Alamar had been working there for more than five years (González Hernández 2005: 126).

This UBPC has done more than just survive and produce. In fact, when I asked Mr. Salcines what he thought his UBPC's most important accomplishments were, he did not mention the impressive production levels but rather the following four areas: first, the increased biodiversity among crops; second, the emphasis on equity and dignified treatment of the co-op members; third, the introduction of scientific principles and new technologies into the work of the UBPC; and fourth, the success achieved in pest control using cultural and biological interventions. Among remaining nagging issues and problems, he listed the difficulties in obtaining sufficient quantities of high-quality organic materials for fertilization and, looming as a future problem, ensuring adequate supplies of water of acceptable quality for irrigation. In any case, this UBPC seems set to continue as one of the main providers of fresh, organically grown produce in the Alamar neighborhood.

CCS Antero Regalades Falcón

My visit to CCS Antero Regalades in the Habana del Este municipality was arranged through FLACSO. Elvira O'Reilly Morris, a graduate student who had recently written a master's thesis at FLACSO on urban agriculture in Havana and had since become the administrator of the CCS Antero Regalades, invited me to visit her cooperative with the approval of MINAG.

This CCS has 51 cooperative members and 17 paid employees. Of the 17 paid employees, 7 are also members without land (*miembros sin tierra*), meaning they have a vote, along with the rest of the members, in decision making at the cooperative. The other 44 members are individual parceleros who hold usufruct plots scattered through the popular council. More than a dozen members raise cattle and produce milk. The rest raise various crops and animals. There is a collective field of 2 ha, also held in usufruct. It contains a 1 ha organopónico that is maintained by two of the collective's employees. Two other employees work at an equine therapy center belonging to one of the co-op members. Most of the remaining employees work in administration or at the CCS's retail points of sale in the popular council.

The cooperative's administrative offices are located in the collective area. The administration contracts with individual members to deliver milk to the state, with the state doing the actual collection of milk from the various farms. The administration also purchases crops from individual members at a 10% discount and markets the produce at the points of sale. Members contribute whatever quantity of produce they can but are under no contractual obligations. In 2006 the CCS produced more than 680 t of vegetables in its collective organopónico and in members' individual parcelas. Because this output is insufficient for local needs, about 40% of the stock at the points of sale comes from other cooperatives in La Habana province by contractual agreement. The prices of these "imports" are controlled by the popular council, but the CCS is free to set the prices of its own members' goods.

The CCS's charter calls for a workers' assembly and a board of directors, who are in charge of making decisions concerning the co-op's activities. The workers' assembly maintains labor discipline through its authority to accept new members, expel existing ones, and terminate the contracts of nonmember workers. Administrative workers work eight-hour days. The workers in the organopónico and the points of sale typically labor longer hours to complete their assigned responsibilities. The legally specified portion of the income generated by the activities of the collective workforce is distributed as personal income to the collective workers. In 2006 this amounted to about 450 pesos per month per worker. Members holding individual parcelas earn only whatever profit they make through their deliveries to the co-op.

Transportation of the produce and tilling of the parcelas present ongoing challenges for members, as only a few members own private cars. They use whatever means of transportation they can get their hands on: ox and horse carts, bicycles, a motorcycle with a side car. The co-op itself does not own a tractor, although occasionally an individual member will rent one.

The collective fields have not experienced pest problems and do not use CREE products. Only lime and tabacina are used in the raised beds. Neem, marigold, and maize serve as repellent and attractant plants and as pest barriers (fig. 7.8). The main difficulty here is water. Basically, the area has had to rely on rainfall and on a small, inadequate well. Some beds cannot be irrigated at all, even when water is available, for lack of 300 m of irrigation line and tubing for its distribution.

The CCS did not have a greenhouse to produce its own seedlings. It could definitely use one, although UBPC Vivero in nearby Alamar is a useful source of seedlings. The organopónico does produce its own

Figure 7.8. Canteros in the collective area of CCS Antero Regalado Falcón. (Photo by author.)

compost, a small quantity of worm humus, and some of its own seeds. The rest of the needed seeds and manure is obtained from a CTA.

As far as professional preparation is concerned, the workers and members here have had access to intensive training provided by ANAP, AC-TAF, and ACPA. More than 90% have attended training sessions. The knowledge base for greenhouse or semiprotected cultivation is already present among co-op members. The only missing ingredient is access to the necessary, scarce—mostly imported—inputs, such as irrigation lines and the structures and irrigation systems needed for semiprotected organopónicos.

One other interesting aspect of the collective area is its close collaboration with area schools. Not only does it provide the schools in the popular council—16 in all—with 2 lb of vegetables per pupil per week, but it also hosts an after-school circle of interest of 10 or so third graders, who spend several hours every Tuesday in the organopónico. In addition, during my visit the workers were in charge of three or four 10- to 12-year-old children who had gotten into trouble at their schools and had been sent to the CCS to do disciplined and disciplinary agricultural work-therapy for one month as a form of school suspension.

Figure 7.9. Organopónico cultivation in Ibraín's parcela. (Photo by author.)

The individual members' parcelas are scattered around the popular council, some at quite a distance from the collective area. Three have huertas intensivas with irrigation. According to Elvira O'Reilly Morris, some parceleros are retirees, while others have jobs in the city and have taken on agricultural work at the parcela as a sideline. All live near their parcelas and come from urban backgrounds. They are quite familiar with marketing and economic thinking, and are well educated and very capable of learning new skills. As a result, CCSs like Antero Regalades tend to be more profitable than the earlier, pre-urban agriculture CCSs were.

We visited several plots belonging to individual co-op members. Right across the street from the CCS headquarters and collective organopónico, and next to a polyclinic, were two adjacent, small parcelas. One, belonging to Ibraín, was a jewel of a small organopónico, growing vegetables— green onions and lettuce, among others—and medicinal plants (fig. 7.9). On the side of the parcela he had a small hut, painted and decorated, announcing itself as the Las Orishas point of sale and selling medicinal plants and herbs used in Santería. Ibraín's parcela benefits from a deal he has with the neighboring polyclinic, allowing him to use for irrigation any excess water from its municipal allocation.

Next to Ibraín's parcela, a few fateful meters removed from the poly-clinic, was a small parcela raising banana plants. Although the bananas were doing well—and were delicious—and lots of chickens were running around the premises, the parcelero's main complaint was lack of water. The spaces between the banana trees were ideal for growing yucca and sweet potato, but the parcelero could not plant them for lack of irrigation water.

In a more rural-looking setting farther away from the collective area, we saw some much larger parcelas belonging to the CCS. Though larger, they were still generally cultivable by a single person. Only three CCS members had plots so large that they needed additional workers. Some parcelas had functioning irrigation, and all seemed to be thriving. Crops raised ranged from radishes, green beans, and okra to beets, onions, and tomatoes, and perhaps others I failed to notice.

But perhaps the most impressive and interesting facility we visited was the Municipal Center for Equine Therapy, one of a network of such cen-ters supported by and collaborating with MINSAP, MINED, and MINAG where children with special needs are offered therapeutic riding, a tech-nique that has found widespread acceptance throughout the world. The owner, Paulino García Rams, has several purebred show horses that he uses in the therapy. The site was full of marabu and rocks at the start of the Special Period, and Paulino and his father had to remove dozens of truckloads of waste in order to start the center. Now, in addition to the horses, they raise rabbits, goats, pheasants, and chickens. They have a contract to deliver eggs and milk to the state through the CCS; practice composting and vermiculture on the premises; and raise parsley, cilantro, oregano, lettuce, chard, onions and green onions, medicinal plants, yucca, and malanga in a small garden. The father is experimenting with breeding a Latin American rodent that can be raised for meat. For all these reasons, GNAU has awarded the facility referencia nacional status (fig. 7.10).

CCSF Arides Estévez Sánchez

This CCS fortalecida, which has a much more prosperous-looking col-lective area than the one at Antero Regalades, lies at the western end of Havana. I visited it only briefly and had a few informal conversations with some co-op administrators and members. The co-op has a collective area

Figure 7.10. Certificate of excellence and other awards on display at the Municipal Center for Equine Therapy. (Photo by author.)

consisting of an organopónico and three greenhouses producing tomatoes and cucumbers for tourist hotels. The collective area is staffed by employees who receive a salary, incentive pay, and a 30% share of any profits. The cooperative itself has a membership of 140 who hold 90 parcelas or fincas; 20 member units are privately owned, and the remaining 70 are held in usufruct. Although the members receive no income from the collective area, they do receive various support services through the CCSF, including technical training, investment loans, and installation of irrigation systems. The co-op members own all the means of production excluding the land. If the co-op were ever dissolved while solvent, each member would receive his share of the value of the assets.

The CCSF produces enough compost and worm humus to make some available for sale to others. The most important obligation of this cooperative is its contracts with the 38 schools it serves, which call for the delivery of 8 lb of vegetables per pupil per month. The CCSF owns a truck with which it makes the deliveries to schools. Other "authorized organizations" also make purchases by contract from the co-op, but they have to pick up the produce at the headquarters and provide their own transportation.

The local government sets the prices in these "for social use" contracts according to monthly or quarterly (for schools) price lists. In addition, the CCSF has multiple points of sale for direct sales to the public (fig. 7.11). The prices at these outlets are not controlled but tend to be set about 20% below the corresponding prices at the free-market mercados agropecuarios, in which the CCS does not participate.

Away from the central area, I visited the small Organopónico Girasol. Two CCSF members hold this plot in usufruct. Only 0.22 ha in size, it has its own well for irrigation and specializes in growing flowers. It produces its own worm humus and has a point of sale right on the premises. It acquires compost from other members of the CCSF who specialize in compost production. The two operators are active as both teachers and students of urban agriculture. One is an irrigation specialist and teaches irrigation to a circle of interest at a nearby elementary school. He states he is not by any means unique: many members of the CCSF either visit schools or receive visits from schoolchildren on their plots. In terms of the co-op members' own continuing education, each Friday at 11 a.m. they

Figure 7.11. The Flamboyán point of sale operated by the CCSF Arides Estévez Sánchez. This CCSF is named after Arides Estévez Sánchez, a hero in the Cuban war in Angola. (The painter evidently misspelled the first name.) (Photo by author.)

all are invited to attend a formal lecture on some urban agricultural topic presented by experts from ACTAF, INIFAT, or MINAG. Also, extension agents visit individual fincas or parcelas to demonstrate new technologies and diffuse successful experiences.

Parcela de Falcón

In Vedado, a formerly upscale Havana neighborhood in the municipality of Plaza de la Revolución, is a parcela of 600 m^2 that colleagues at FLACSO arranged for me to visit. Back in 1992 the site was an abandoned lot overgrown with weeds and being used by the population as a wildcat dump for urban refuse. A. Falcón, an ordinary urban worker with no background in agriculture, was given the task of rehabilitating this site in order to begin producing medicinal plants.[3] At that time, medicinal plants were beginning to become important in the Cuban health system, as the government turned to herbal medicine in the import-starved atmosphere of the Special Period.

Falcón's first move was to immerse himself in libraries to read the existing literature about raising plants in an urban environment, including works about composting and vermiculture, and scientific works about medicinal plants. At the same time he undertook the arduous task of cleaning up the site. Fifteen years later, in 2007, more than 40 different medicinal plants and herbs were being grown here in organoponic-type beds. The substrate utilizes compost and worm humus obtained from waste vegetable matter and manure obtained from chickens raised and kept there for egg production. Concerning the worm humus, Falcón says, "I started with a few hundred worms, now I have millions." For his compost pile he uses waste generated at the parcela, plus on weekends he collects fallen leaves from the neighborhood streets with a wheelbarrow and obtains orange peels and other refuse from the farmers' markets. Thus, all of the organic material for his organopónico beds is produced right here at the parcela. He has plans to start raising rabbits and tilapia (in fish tanks) with the aim of producing yet more organic material for fertilization. For example, he intends to use water from the fish tanks, which is rich in organic material, for irrigating the crops.

Falcón uses neither CREE products nor chemicals in his parcela. Pests are controlled by diversification and repellent plants. When snails invaded the plot, he gathered them off the plants manually, killed them, and threw the remains on the compost pile after making sure they were dead. There are, in general, few problems. If one arises that Falcón cannot handle on his own, he seeks advice and help from the local CTA. He also now has the support of three additional workers who labor at the parcela. His only transportation, however, has remained one old, beat-up bicycle! All of his sales occur right at the parcela: neighbors come there to buy medicinal and "spiritual" (that is, for Santería) plants at 2 pesos a bunch. More important than Falcón's role as a producer of medicinal plants, however, is his willingness to share with the community the vast knowledge he has acquired over the years. He welcomes groups of students sent on field trips to the parcela. He visits schools to give workshops on growing and preserving herbs and plants, permaculture, soil sustainability, and rooftop and home gardens. He advises neighbors on how to start their own garden or rooftop production, and on how to preserve what they harvest. He even shared with me one of his recipes for a condiment of dried herbs and leaves: 30% oregano; 30% basil; 20% onion, green onion or mountain garlic; 10% leaves of bitter orange; and 10% guava leaves. He also consults with medical doctors, many of whom do not have a strong background in herbal medicine.

For Falcón, the strongest incentives are intrinsic ones. He is concerned about the future of our planet and about what kind of world we are going to leave to our children. He is gratified when he is able to contribute to healing someone who is sick. His commitment to improving the environment of his community has also led him to participate actively in Mi Programa Verde (My Green Program), an ACTAF tree-planting initiative in the city. He has planted ornamental plants, many rescued from among plants other people had discarded, along the 10- to 15-yard-long pathway leading from the street to the gate of the parcela. And between the sidewalk and the curb of the street in front of the parcela, he has planted several trees. What used to be an abandoned lot full of refuse is now green space, and Mi Programa Verde is thriving here. Falcón told me three driving forces enable success in any action in life, even under difficult circumstances: necessity, possibility, and will. As long as the object is possible to achieve, necessity is the mother of invention, and will to succeed is crucial to achieving it. This attitude undoubtedly explains how a parcela like Falcón's came to be.

La Hierba Buena

Another interesting individual contribution to urban agriculture is the patio of Dr. Raúl Gil Sánchez, at his house in the popular council of Villa, municipio of Guanabacoa, Havana. Dr. Gil is a professor at the Instituto Superior de Ciencias Médicas, has a master of science degree in social psychiatry, and is the director of a mental health center in the municipio of Regla in Havana. After becoming interested in the efficacy of herbal therapy for treating mental health problems, he decided to learn how to grow plants for use in his clinic. He began by contacting Miguel Salcines of the UBPC Alamar, which had recently been organized, for help. It was a propitious contact. At the time of my visit, his garden was home to hundreds of plants; and Salcines has continued to give advice and help, ranging from the still-pending contribution of worms for vermiculture to technical support in growing bonsai trees.

Back in 1995, Dr. Gil's backyard was neglected and barren, with hard ground. Neighboring his house was a vacant lot, once a pottery workshop, a hostel, and a warehouse but fast becoming an eyesore and a garbage dump, infested with rats and cockroaches and the site of undesirable, even criminal activities. Dr. Gil asked the local authorities to let him take charge of the site in usufruct, with the promise that he would clean it up and put it to good use. When his request was granted, that is exactly what he proceeded to do. Over time, he and his family and friends hauled away 23 truckloads of trash from the lot. He then combined his backyard and the usufruct plot into a single unit measuring 437 m², a little more than one-tenth of an acre. And he turned it into a thriving garden named La Hierba Buena (The Mint).

In 2000, La Hierba Buena was officially registered as an urban agricultural patio, following inspection by three representatives from MINAG. This opened the door to technical assistance from MINAG and extension agents at the local CTA. Seeds, other inputs, and technical assistance come from the CTA. MINAG provides free organic material. It also organizes networks among various producers so that they can share their experiences.

In 2007 the thriving garden was home to lettuce, beets, peppers, green beans, spinach, onions, Swiss chard, parsley, cilantro, medicinal herbs and, very appropriately, mint—growing in walled beds constructed with cinder blocks and in small beds constructed in what appear to be large

Figure 7.12. La Hierba Buena patio, belonging to Dr. Raúl Gil Sánchez.
(Photo by author.)

water pipes cut in half—as well as banana, mango, tamarind, mandarin,
fig, guava, lemon, and anon trees (fig. 7.12). To plant the trees, Dr. Gil
excavated holes 1 m in diameter and 80 cm deep in the hard ground, then
refilled them with a mixture of soil, compost, and worm humus before
transplanting seedlings in them. Oregano (a pest repellent) and maize
(which attracts beneficial insects) are grown for pest control. Neem and
noni trees provide additional barriers. Dr. Gil plans to start producing his
own worm humus in an old bathtub in the garden. In a second bathtub

he raises frogs for mosquito control. A one-page brochure on La Hierba Buena describes current (in 2007) activities in the patio: the production of vegetables, fruits, ornamental and medicinal plants, the presence of a Martí garden (containing flora described in the journal of the Cuban patriot), a cultural circle, and a group for human growth. Future plans, in addition to producing worm humus, are to increase the production of vegetables and to introduce aquiculture. And the following quotation from José Martí is highlighted: "To live on the land is nothing but the duty to do it good" (my translation).

To keep a garden like this going, Dr. Gil manages to carve out two hours each weekday from his hectic professional life to work in the garden along with his wife. Saturday and Sunday mornings are also spent in the garden. Yet this patio does not produce anything for sale. Like most of the more than 60,000 patios in Havana, all produce is used for self-sufficiency or shared with neighbors. According to Dr. Gil, La Hierba Buena's task is to open up and create alternative spaces for sociocultural development within the community, to educate local children in ecology, and to contribute to the formation of sound values among youth. Thus its contribution to the urban agriculture of Havana lies not so much in production as in the physical and social environment of the community.

And this impact is substantial. Every Saturday morning, Dr. Gil and his wife host an educational workshop in the patio for neighborhood children. During my visit, which happened to be on a Saturday, about a dozen youngsters, mostly under the age of 10 and including the Gils' own children, were in attendance. The first half of the workshop consisted of theoretical instruction and discussion of a physical or social-environmental issue. The children learned how to plant seeds and grow plants, how to be sensitive to the environment, and how to treat each other, as well as social skills not directly related to agriculture, such as how to behave at the beach. They also engaged in creative activities such as drawing pictures. This was followed, after a break, by actual gardening. Dr. Gil views these children's workshops as environmentally focused community development: at the same time food is produced, attitudes concerning the physical and social environment are also changed.

Dr. Gil's patio first drew national attention when a paper he wrote for a meeting of local urban agriculturalists was chosen to be presented at a national scientific congress and won a national prize. This paper described the process of cleaning up and rehabilitating neglected urban sites through urban agriculture. Further national recognition came when GNAU awarded La Hierba Buena referencia nacional status. This status confers some additional material benefits on the patio. For example, technical assistance is available from the CTA at heavily subsidized prices compared to their normal fees. Of course, along with national recognition comes national scrutiny and periodic visits by supervisory teams from GNAU to reaffirm its referencia status.

Conclusion

The case studies presented in this chapter are snapshots of a process Cuba has been undergoing since the 1990s. They illustrate how the support systems described in the first six chapters of this book—which were put into place by central authorities with associated patterns of social resource allocation—combined with Cuba's ample human capacity and the new social consciousness introduced after the revolution to enable an agricultural transformation. Cubans in perhaps unexpected places and from definitely unexpected backgrounds stepped forward to take on the tasks of producing wholesome food, engendering community, and cleaning up and beautifying the environment.

Clearly, a disproportionate share of the production units I was able to visit were places that had distinguished themselves and achieved some level of referencia or excelencia status. On the other hand, none of the sites I visited in Matanzas or Pinar del Río were of referencia nacional or candidate to such a status (although San Cristóbal had aspirations in that direction). Also, of the more than 50 members of CCS Antero Regalades in Habana del Este, only Paulino García and his equine therapy center had been singled out as a unit of excellence. The rest were struggling usufruct holders trying to make a go of it under somewhat adverse circumstances. In the next chapter I will turn from individual cases to an assessment of the overall results of urban agriculture in Cuba.

8

◇◇◇◇◇◇◇◇◇◇◇◇◇◇◇

Evaluating the Success of Cuban Urban Agriculture

Urban agriculture in Cuba has had a significant impact in several areas of the society. Any overall assessment of the shift to urban agriculture has to pay attention to all of these impacts. This chapter considers four of the most important: food production, employment, environmental protection, and community building. Because food production is obviously the most important, the discussion starts here.

Food Production and Distribution

Within the fold of urban agriculture, the subprogram of Vegetables and Fresh Condiments stands out in two respects. Among the 28 subprograms, it has the most extensive, publicly available data, and it represents urban agriculture's most successful and consequential effort in contributing to national food security. Its story, in terms of its historical development and current achievements, deserves to be told in some detail.

Food Production

Table 8.1 displays the output of this subprogram since its inception at the macro (all of Cuba), meso (Havana), and micro (UBPC Alamar) levels. The increases in output are spectacular at all three levels. The thousand-fold increase in Cuba as a whole between 1994 and 2006 corresponds to an annual growth rate of 78%. This kind of growth obviously required a large increase in cultivated area, but yields per square meter also went up impressively, especially in the early years of the introduction of organopónicos (Rodríguez Castellón 2003).

Table 8.1. Production of Vegetables and Fresh Condiments subprogram in Cuba, Ciudad de la Habana province, and UBPC Organopónico Vivero Alamar

Year	Cuba (1,000 t)	Ciudad de la Habana (1,000 t)	UBPC Alamar (t)
1994	4.2	—	—
1995	16.0	—	—
1996	58.0	—	—
1997	140.0	20.7	20
1998	480.0	49.9	25
1999	876.0	62.2	23
2000	1,680.0	120.1	43
2001	2,360.1	132.2	45
2002	3,345.0	188.6	75
2003	3,931.2	253.8	108
2004	4,194.8	264.9	113
2005	4,074.5	272.0	143
2006	4,200.0	275.0	235

Sources: Acosta Mirrelles 2006a; Rodríguez Nodals, Companioni Concepción, and González Bayón 2006; Salcines López and Salcines Milla 2007.

Between 2001 and 2006, although output continued to increase substantially—by 77% in Cuba and 46% in Havana—average yields per unit actually declined from 12.7 kg/m^2 to 7.6 kg/m^2. This decrease in yields, in fact, has a rather benign explanation. There are four different modalities of cultivation in the Vegetables and Fresh Condiments subprogram: patios, parcelas, huertas intensivas, and organopónicos. The average yields realized in 2006 in these four modalities were, in order of increasing efficiency, 5.79 kg/m^2 in patios, 6–8 kg/m^2 in parcelas, 11.3 kg/m^2 in huertas intensivas, and 18.44 kg/m^2 in organopónicos (Rodríguez Nodals, Companioni Concepción, and González Bayón 2006). Since 2001 the modalities associated with lower yields, namely patios and parcelas, have expanded dramatically. Thus, in 2001 there were under cultivation in this subprogram of urban agriculture a total of 18,591 ha, with 13,966 ha of patios and parcelas, 3,953 ha of huertas intensivas, and 732 ha of organopónicos (Rodríguez Castellón 2003). By 2006 the total cultivated area had expanded from 18,591 to 52,389 ha: 12,774 ha in patios and 30,875 ha in parcelas (43,649 ha total), 7,557 ha in huertas intensivas, and 1,183 ha in organopónicos. As is apparent, the area cultivated increased in all modalities, but particularly in those with the smallest yields: there were 61% and 91% increases, respectively, in organopónicos and huertas intensivas, compared with a 212% jump in patios and parcelas.

The increase in the number of patios is largely due to rapidly increasing official registration of patios, as GNAU attempts to incorporate all home cultivation into the official urban agriculture program. In 2007, GNAU estimated there were 1.5 million home gardens in Cuba, fewer than one-third of which are registered.[1] What portion of the increase in patios is due to families deciding to use previously barren space for new gardens versus deciding to register preexisting patios is unknown, but surely the vast majority of the increase is due simply to registration.[2] The benefits of such incorporation seem obvious: better supervision and problem detection, better horizontal communication among producers, enhanced biodiversity, better control of plant diseases, and improved provision of extension services to otherwise isolated producers.

Although Havana was the main focus of innovation and effort early on, since it had the greatest need, the other provinces have progressed rapidly in recent years. Havana's share of the total output of the Vegetables and Fresh Condiments subprogram dropped from about 35% in 1998 to a mere 6.5% in 2006, and it ranks lowest in terms of output per capita.[3] In fact, one of the cardinal principles of the urban agricultural effort in Cuba has been the insistence that development take place as uniformly as possible across the urban spaces of the island as well as across all of the 28 subprograms. The data in table 8.2 document the successful application of this principle. Whereas the population ratio of the largest to the smallest province is 5.4, the corresponding spread between the maximum and minimum provincial production levels in this subprogram is only 2.2.

Information I was given during my visit in Matanzas indicates that the same kind of uniformity obtains at the municipal level. Matanzas has 14 municipalities, two of which have unique characteristics. Ciénaga de Zapata is the largest municipality in area but is a sparsely populated swamp along the southern coast of the island and is almost entirely a nature preserve. The other is perhaps the best-known Cuban municipality in the world: Varadero. This small municipality along a coastal strip is home to the most famous tourist beaches of Cuba and to dozens of four- and five-star hotels. These two municipalities each have less than 20 ha of land in urban agricultural cultivation and are not significant producers. If one excludes these two outliers, the other 12 municipalities all have ample urban agricultural activity taking place within their boundaries. Matanzas, the most urbanized municipality with the largest population, is not the biggest urban agricultural producer. That honor goes to the municipality

Table 8.2. Distribution by province of total and per capita production in the Vegetables and Fresh Condiments subprogram of urban agriculture, 2001 versus 2007

Province	Population, 2001	Total production, 2001 (1,000 t)	Daily per capita production, 2001 (g)	Population, 2007	Total planned production, 2007 (1,000 t)	Daily per capita production, 2007 (g)
Pinar del Río	737,342	183.2	681	731,232	364.1	1,364
La Habana	707,764	182.3	706	739,967	445.4	1,649
Ciudad de la Habana	2,186,632	136.6	171	2,156,650	283.8	361
Matanzas	661,901	145.7	603	684,319	249.1	997
Villa Clara	836,322	168.2	551	809,231	306.0	1,036
Cienfuegos	396,691	182.3	1,259	402,061	290.2	1,997
Sancti Spíritus	462,320	179.4	1,063	464,221	209.7	1,238
Ciego de Ávila	410,701	171.5	1,144	420,996	300.5	1,956
Camagüey	789,883	181.7	631	783,372	336.1	1,175
Las Tunas	530,328	161.4	834	533,127	204.5	1,051
Holguín	1,032,670	184.4	489	1,035,744	300.8	796
Granma	832,644	164.7	542	833,600	375.5	1,234
Santiago de Cuba	1,037,690	157.2	415	1,044,698	327.5	859
Guantánamo	514,121	145.6	776	511,063	292.5	1,568
Isla de la Juventud	80,091	15.1	520	86,509	14.0	443

Sources: GNAU 2007a; Rodríguez Castellón 2003.

of Jagüey Grande, but it produces only 3.2 times the output of the least-producing municipality of Martí (GNAU de Matanzas 2007).

The rest of the food-growing urban agricultural subprograms are less advanced than Vegetables and Fresh Condiments, yet all have begun making, albeit at times modest, contributions to the nutrition of the Cuban population. Available GNAU projections for 2010 per capita production in various subcategories of these subprograms follow:

Dried Herbs	0.149 kg
Fruits	63.2 kg
Rice	6.5 kg
Coffee	0.255 kg
Bananas (fruit and cooking)	2.7 kg
Roots and Tubers	54 kg
Beans	
cold season	2.4 kg
summer	0.54 kg
Maize	4.02 kg, or about 11 ears
Sorghum	0.76 kg
Animal feed (silage and fodder)	75.6 kg
Poultry	0.97 kg
Rabbit meat	0.23 kg
Sheep meat	1.89 kg
Goat meat	0.73 kg
Small-scale food processing	6.2 kg

In addition, about 20 million liters of goat milk and 169,000 liters of vegetable oil are to be produced, the latter obtained from peanuts, sunflowers, sesame seeds, and soybeans. Egg production will reach 540 million eggs, or about four dozen eggs per person. There are to be 155,720 beehives producing honey and pollinating 49,762 ha. Aquiculture will be practiced on 12.92 ha of stillwater locations.[4] Because it is impractical to raise large livestock such as cows and pigs in urban locations, the *Lineamientos* do not estimate production levels for beef, cow's milk, and pork in urban agriculture. It has been estimated, however, that urban and peri-urban small-scale pork production had reached 12,000 t annually by 2008, corresponding to 1.08 kg/person/year (Grogg 2008a).

Food Distribution

Of equal or even more pressing importance than food production is the timely distribution of this output for consumption by the population. Given the scarcity of fuel and transportation infrastructure in the Special Period, keeping the producer-to-consumer chain short has been a crucial priority in urban agriculture. Close to 60% of the vegetables grown in urban agriculture are sold directly to the public at stands situated at the production unit. The UBPC, CCS, and CPA cooperatives also maintain more than 10,000 points of sale at other locations nearby (Armengol 2008; Ministerio de Relaciones Exteriores 2008). As table 8.3 depicts, no other distribution channel even comes close to the category of direct sales. Production in patios, primarily destined for consumption by the family, and self-provisioning gardens of workers in their workplaces account for an additional 21% of the output. Finally, about 1% and 10%, respectively, end up being marketed in MALs, managed by the Ministry of Domestic Trade with prices determined by the free interplay of supply and demand, and in MAEs, managed by MINAG with controlled prices that fluctuate in tandem with MAL prices. In other words, the "free" agricultural markets

Table 8.3. Distribution of vegetables and fresh condiments produced in urban agriculture, 2004

Destination	Metric tons	Pct. of total output
Direct sales to the public	2,492,523	59.41
State markets (MAEs)	419,612	10.0
Schools	60,102	1.43
Day-care centers	6,149	0.15
Hospitals	7,794	0.2
Other social services	12,386	0.29
Universities	15,461	0.37
"Free" markets (MALs)	41,948	1.00
Worker self-provisioning at their workplaces	293,639	7.0
Post-harvest losses	96,481	2.4
Small agro-industry	82,117	2.0
Family consumption (patios)	647,697	15.4
Other uses	4,146	0.1
Animal feed (residues)	20,971	0.5

Sources: Companioni Concepción 2006; Rodríguez Nodals, Companioni Concepción, and González Bayón 2006.

play a very modest role among the outlets marketing urban agricultural products, and even the network of MAEs is not substantial. Once projected to contain about 900 markets, the MAE network has consolidated to 550 markets in all of Cuba, 310 of them in Havana (Martín González and Pérez Sáez 2008; Cabrera Balbi 2009). In terms of quality, assortment, and availability of products, the MAE markets in general have not been able to match the standard of the MAL markets. On the other hand, the prices are 20%–30% below MAL prices and are comparable to the prices at the direct points of sale.

The three main channels through which urban agricultural output reaches the final consumers—direct sales, self-provisioning, and state MAEs—distribute more than 90% of the vegetables and fresh condiments harvested. Of the rest, about 2.4% is lost through post-harvest spoilage. Another 2% is processed and conserved through small agro-industrial activity. The rest, about 2.5%, destined to various social ends, is deployed in many directions, including, for example, support for the Sistema de Atención a las Familias (System for Attention to Families), which provides meals for homebound elderly, deliveries to maternal homes, and provisions for some military units. The last category is quantitatively small, as the military is mostly self-provisioning; the total urban agricultural support for military units amounts to less than two-thirds of the distribution to day-care centers. Deliveries also go to hospitals, MININT facilities, and vegetarian restaurants, among other sites.

The largest and most ubiquitous social use of urban agriculture is, however, the education sector. Of the roughly 105,000 t per year destined for communal use, almost 82,000 t go to day-care centers, schools, and universities. Although this represents only about 2% of the urban agricultural output, the amounts are sufficient to make a very substantial contribution to the diets of preschool and school-age children. Since 2001, when MINAG decided to charge the urban agriculture movement with responsibility for supplying vegetables to schools, the program has expanded from an initial set of 60 prioritized centers—35 schools and 25 day-care centers—to cover the entire island. Supplies are arranged through direct contracts between schools and individual production units in the vicinity, at below-market prices set by the government, with the producers being responsible for actually delivering the produce to the school. The delivery goal was set at 187 g of vegetables per day per student, thus ensuring that students would consume at school more than half of the 300 g per day set as a minimum by the FAO. The short supply chains from production unit to school ensure the vegetables are fresh at the time of consumption—perhaps increasing their acceptability to picky eaters (Puente Nápoles 2006). Given the generally low use of vegetables in the traditional Cuban diet, the program included deliberate efforts to educate teachers, students, and, yes, the cooks about the benefits of eating the vegetables being delivered to their school doors.

A few simple calculations suffice to judge the adequacy of the total deliveries. Cuban primary and secondary students receive about 1,000 hours of instruction per year (Gasperini 2000: 29–30). Assuming this corresponds to about 200 days of schooling per year, each of the 1.8 million students would receive about 170 g per day. In the same year, day-care centers for children under age six had an average daily attendance of 110,000 (Oficina Nacional de Estadísticas 2009, table 18.3). The 6,149 t of vegetables delivered to these centers would provide 280 g per child per day under the same assumption of 200 annual days or, assuming that the day care operates six days per week year-round, about 180 g per child per day. Preschool children thus seem more amply supplied with vegetables than their school-age counterparts.

That urban agricultural output has made a significant difference in the nutrition of Cubans seems obvious. Even determined critics of Cuba's overall agricultural performance appreciate the most compelling successes of urban agriculture (Espinosa Chepe 2006: 14–15). As table 8.2 shows, by 2007 all Cuban provinces met the FAO guideline of providing at least 300 g of vegetables per person per day. In fact, all but Ciudad de la Habana province already did so by 2001. It would seem that Cubans are enjoying healthier diets because they are consuming more vegetables, and the vegetables themselves are ecologically grown. In terms of the percentage of overall nutrients supplied by urban agriculture, the results are not as spectacular. Nova González (2006: 283) gives the following data for 2005. By that year the per capita daily caloric intake of the Cuban population averaged 3,356 calories, signaling a complete recovery in this regard from the disastrous food crisis of the 1990s, which had reduced daily caloric intake to as low as 1,863. Thus, by 2005 and as part of its more general economic recovery, Cuba had once again reached and exceeded the pre-crisis levels of calorie consumption. Notably, however, the bulk of this consumption—nearly 60% of the calories and somewhat more than 60% of the protein consumed—came from sources other than urban agriculture, which rely heavily on imported foods.[5] It can be concluded that although comprehensive food sovereignty continues to elude Cuba's grasp, urban agriculture has ensured that food security and sovereignty have been achieved in the area of vegetable production.

The introduction of urban agriculture has other physical and social benefits that are equally, if not more, significant in the long run for Cuban society than the provision of foodstuffs. These benefits affect the work and communal lives of the Cuban population, as well as the physical conditions under which they live. They may be broadly grouped and discussed under the categories of employment, environment, and community. If the production and employment effects can be thought of as countering the crisis that followed the collapse of the previous paradigm of agriculture, the other two are more associated with the healing of ills resulting from that previous paradigm, even when it was functioning well in its own terms.

Employment

In the early 1990s, the economic collapse of the Soviet Union and COMECON caused the Cuban economy to implode. The virtual ending of international trade—especially of the sugar-petroleum exchange with the Soviet Union—had obvious consequences for the Cuban economy. The resulting cessation of economic activity—especially in industry and transportation—led to widespread unemployment as factories and buses stopped operating due to lack of fuel and other inputs. GDP declined by more than 30%. Unemployment skyrocketed in the cities—although it was a Cuban-style, relatively humane form of unemployment, as workers kept their affiliations with their old workplaces and 70% of their salaries (Koont 1994). The new employment opportunities in the cities opened by urban agriculture offered a very welcome shot in the arm to the ailing urban economy.

By 2001, more than 208,000 workers in Cuba (Sanz Medina 2001) and more than 20,000 in Havana (Rodríguez Castellón 2003) had found employment as urban agriculturalists. By 2006 these figures had risen to more than 350,000 for Cuba and 44,000 for Havana (Pagés 2006b). These numbers count only those workers whose primary employment is in agriculture, that is, members of UBPCs and CCSs, parceleros, and workers in state-owned urban agricultural units. It does not include the estimated 1.5

million home gardeners who grow food to feed their families (Rodríguez Nodals and Companioni Concepción 2006).

Not only has employment in urban agriculture grown rapidly, but the composition of this agricultural workforce is significantly different from the traditional rural agricultural workforce. As late as 1996, labor force participation rates were only 25% for rural women, 43% for urban women, and 70% for men in both rural and urban settings (Cuba 2002). In rural Cuba, where fieldwork was still considered a man's job, proportionally many fewer women participated in the workforce. The shift to urban agriculture has increased participation by women for at least three reasons. One natural reason is that labor force participation rates for urban women are simply higher. Another is that the scientific and technical sophistication of urban agriculture has raised the demand for (and thus participation of) skilled labor. Women have used their considerable edge in formal education to land skilled positions that require professional skills in administration, technology, and science. According to data compiled by Adele Cuba (2002), women make up more than 50% of the skilled (mid-level technical or university-educated) workforce in Havana's non-state agricultural sector. In addition, there have been concerted efforts and campaigns to attract more women into urban agriculture as workers. The Federación de Mujeres Cubanas (FMC, Federation of Cuban Women), for instance, has organized a large organopónico called Las Marianas, employing 140 women field-workers (Murphy 1998: 17). The FMC has also organized 174 Comités Femeninos (Women's Committees) where issues and problems that arise for female participants in this sector, and proposals for their resolution, are discussed. As a result of these efforts, fully a quarter of the workforce in urban agriculture in the City of Havana province is now female.

One of the inducements attracting workers to urban agriculture is a remuneration structure that yields incomes substantially above those for most state workers. Rodríguez Castellón (2003) asserts that workers who switch their employment to urban agriculture can improve their incomes by 10%–20%. Thus, not only has urban agriculture generated jobs, but those jobs are at the upper end of the pay scale and in substantial numbers.

Due to the nature of the technologies it employs, urban agriculture is inherently quite labor intensive. Small-scale, bio-diverse, agroecological cultivation renders use of machinery and automation, and the application of chemicals for pest control and fertilization, impractical. Instead, the management of crops is once again returned to the meticulous attention of human beings. All phases of this care—from soil preparation and conservation to sowing and harvesting, fertilization, and irrigation—require skilled labor. A 25-acre farm employing 150 workers would be unthinkable in the United States or Europe, but Miguel Salcines's UBPC in Alamar does exactly this, while generating quite attractive incomes for its members.

This labor-intensive paradigm of crop-raising has led to the employment of some 350,000 workers, or 7% of the entire 5-million-person Cuban workforce. One in 14 of all Cuban workers, and 1 in 3 workers in the agricultural sector, now labors in urban agriculture (Oficina Nacional de Estadísticas 2009, table 3.4). By the middle of the first decade of the twenty-first century, the high urban unemployment of the Special Period had been largely eradicated, thanks in no small part to the new urban agriculture paradigm.

The Environment

Urban agriculture can be considered a program of environmental protection and preservation as well as a program of food production. Ever since the triumph of the revolution, Cuba has been keenly aware of local and global environmental issues. The government has passed many environmental laws establishing nature preserves, protecting forests and lakes, and so on, and providing for administrative enforcement of such laws. It even established a ministry, CITMA, that combines science, technology, and the environment in its very name. Even in the context of all of this environmentally focused activity, urban agriculture stands out as one of the most important environmental initiatives in Cuba, and is so perceived by Cuban leaders. Evidence for this assertion can be found in the book *My Life* (published in Spanish as *Cien horas con Fidel*), based on 100 hours of interviews with Fidel Castro by Ignacio Ramonet, editor of *Le Monde Diplomatique*. When, near the end of the series of interviews, Ramonet asked Castro about the most important Cuban initiatives to protect the environment, Castro did not discuss environmental laws and regulations, but rather cited urban agriculture (Castro Ruz and Ramonet 2008: 400).

There are a couple of good rationales for Castro's response. First is the general effect of shifting food production away from reliance on fossil fuels and petrochemicals. Conventional industrial agriculture emits large quantities of carbon dioxide, associated with global warming; degrades soil and water; and has toxic effects on human beings, both producers and consumers. The emergence of agroecological urban agriculture in Cuba has brought with it a new level of care and concern for the quality of the soil and water and of the produce itself, leading to a generally healthier environment.

Second, urban agriculture participates actively in Mi Programa Verde (My Green Program), the tree-planting initiative organized and championed by ACTAF in both urban and rural Cuba since 1996. This program for the "greening" of Cuba extends the effort that began in 1959 to reverse the deforestation Cuba had been undergoing throughout its colonial and postcolonial history. By 1959, only 14% of the island's surface area was still covered by trees. By 2006, Cuba had managed to extend its forested lands to 23.6% of the surface area. In 2007, as part of a UN-sponsored, worldwide push for reforestation, Cuba planted 136 million saplings, 2.1 million of them in Havana (Grogg 2008b). The goal is that by 2020, the forestation of Cuba's surface area will rise to nearly 30%. Havana, for its part, already has 23 m^2 of green space per inhabitant, exceeding the World Health Organization guideline of at least 10 m^2 of green area per person (Peláez 2009). The province is currently well on its way to increasing its forested areas to more than 10% of its total surface area (Sierra 2009g).

In the cities, Mi Programa Verde is conceived as a project of, by, and for the community that aims to cover all appropriate available spaces with ornamental, fruit, and forest-species trees. Among its goals are several clearly urban agricultural ones: the growing of food for humans and feed for animals, and the production of organic materials to supply the compost piles and produce fertilizers for urban agriculture (ACTAF 2006b). The urban agriculture movement guided by GNAU intersects with Mi Programa Verde in at least three subprograms: Fruit Trees; Ornamental Plants and Flowers; and Forest, Coffee, and Cacao Trees. According to a recent survey, Havana possesses about 138,000 mature trees in its gardens, parks, public pathways, and streets. The plan in the *Lineamientos* for the Forest, Coffee, and Cacao Trees subprogram calls for planting more trees in all the provinces. In Havana the goal was to double the trees of this type from 15,000 in 2008 to 30,000 in 2010. The corresponding increase for all of Cuba was from 169,000 to 338,000 (GNAU 2007a: 39).

The benefits of urban reforestation effort are manifold. One, of course, is increased availability of fruits, coffee, and cacao, as well as wood for furniture-making and construction. Not to be ignored are secondary benefits such as urban beautification, physical and psychological well-being of the population, reduction in urban stress, increased privacy for households, screening or better yet replacing of urban eyesores, enhancement of biodiversity, and reduction of air and noise pollution through the filtering action of plants. These environmental consequences certainly add considerably to the overall value of the contribution of urban agriculture to Cuban society and should not be ignored in any overall assessment of the program.

Community Building

The shift to urban agriculture in Cuban cities both enabled and necessitated efforts in community building and organization. The small-scale, local, labor-intensive techniques employed require involvement of community members not only as consumers, but also as producers. Dietary changes to include more vegetables, some of them barely known in Cuba previously, contributed to communal health, but also required community education on nutrition.[6] Furthermore, the urban agricultural workforce had to be recruited locally. This need to recruit and involve the community is clearly reflected in the way urban agriculture operates. In fact, both by design and intention and in consequence, community organizing and community building have been at the very core of urban agriculture. The promotional and educational literature of and about urban agriculture—whether published by GNAU or ACTAF or appearing in *Agricultura Orgánica*—is replete with references to "de la comunidad (or barrio)," "por la comunidad," and "para la comunidad" (of, by, and for the community or neighborhood).

When urban agriculture supplies vegetables to day-care centers and schools, it is contributing to community health by introducing and advocating nutritional changes associated with healthier diets. When an urban agricultural unit sponsors one of the 3,000-plus circles of interest in the primary and secondary schools of urban Cuba, it is promoting the formation of future workers knowledgeable in urban agricultural technologies and ensuring the local community's continuing attachment to the urban agriculture project. When a unit takes under its wing a youngster who has gotten into trouble at school and provides work- and nature-based therapy, it is contributing to community healing.

When vacant lots that have turned into garbage dumps are reconstituted as green spaces full of trees, flowers, vegetables, and ornamental plants, the result is not only aesthetic improvement—although urban beautification is certainly a desirable end in itself. In urban agriculture, however, these sites typically also become focal points in the community. Whereas formerly they avoided the unsightly unofficial garbage dump and site of criminal activity, people now come to the new neighborhood parcela or organopónico to buy vegetables, fruits, and medicinal and spiritual plants and to interact with their neighbors. The difference such a change makes in the sense of community well-being and cohesiveness is bound to be substantial.

GNAU is keenly aware of these community-focused aspects of urban agriculture, as is clear from the *Lineamientos*. The discussion in the chapter on the Forest, Coffee, and Cacao Trees subprogram asserts that family participation in planting a tree in the yard or home garden is important because it symbolizes the strength of family ties, strengthens feelings of community solidarity, and raises the interest of the next generation in the ongoing reforestation activities (GNAU 2007a: 37). The *Lineamientos* also assert that growing flowers and ornamental plants in urban agricultural production units wherever feasible will nourish the spiritual needs of the community. Giving flowers to a mother, sister, girlfriend, or even coworker is an indispensable part of everyday life and a practical necessity. An explicit part of the mission of this subprogram is to support this aspect of community life, promoting neighborhood harmony. Urban agriculture planned to produce 155 million dozens of flowers in 2010 (GNAU 2007a: 29–30).

As this chapter has demonstrated, the paradigmatic shift to urban agriculture in Cuba has had significant consequences beyond the production of food. This fact ensures that any future discussions about the continued viability and desirability of this new paradigm must involve much more than just questions of food production. The implications of this situation for the long-term prospects of urban agriculture in Cuba need to be assessed carefully. That is the topic of the following chapter.

9

◇◇◇◇◇◇◇◇◇◇◇◇◇◇

Looking to the Future of Urban
and Sustainable Agriculture

Cuba and Worldwide

By the time Cuba was forced into growing much of its domestic food supplies within cities, countries worldwide had started paying serious attention to issues of urban and sustainable agriculture. This final chapter uses a wider lens to examine not only where Cuban urban agriculture is headed but also recent urban agricultural efforts in Latin America and the third world. Then, in conclusion, it refocuses to a more evaluative level, examining the costly social and environmental consequences of the current agricultural paradigm of rural, industrialized monoculture. To what extent can we apply the lessons of the Cuban experience to the rest of the world, and what would be the benefits and challenges of the transition to a more small-scale, ecologically friendly model of food production?

Whither Goes Cuban Urban Agriculture?

Even in the context of continuing, and even worsening, difficulties in the country's agricultural performance (Hagelberg and Alvarez 2007), urban agriculture has been rapidly advancing and becoming consolidated in Cuban society. It still faces internal challenges and constraints from several sources, however. First, there is historical inertia among the older generations of agronomists and Ministry of Agriculture officials who thrived much more comfortably in the era of Soviet-style industrial agriculture. Some of them surely still consider the agroecological adaptations of the recent decades to be a crisis-forced, temporary phenomenon and would welcome back labor-reducing machinery and yield-raising chemical fertilizers and pesticides, once the easing of the economic crisis made these options possible once again. A recent case-study-based article found that small farmers in Cuba showed little evidence of a philosophical conversion to the paradigm of agroecological production (Nelson et al. 2008). The authors interviewed farmers in the municipality of San José de las Lajas, La Habana province; most were members of rural CCSs and CPAs, but several cultivated units classified as urban agriculture.[1] These farmers by and large had a pragmatic attitude toward the paradigm shift to agroecology: it was forced on them by nonavailability of the necessary resources for conventional agriculture, and they would readily revert to using chemical and fossil-fuel inputs if and when they became available again. Yet, even among the farmers who yearned to return to the "good old days," there was an acknowledgment of the benefits of agroecology and a general respect for the ideas of reducing use of "poisonous agrochemicals" and of increasing biodiversity. They were also thankful for the improvement in soil quality that resulted from organic fertilizers and the removal of soil-compacting tractors from the fields. At the same time, Nelson et al. found that the desire to return to familiar ways was strongest among people who were agronomists or farmers before the Special Period. The newer recruits to the urban agriculture movement were much more likely to be ideological converts to agroecology, as their entire training had taken place within the agroecological paradigm.

Although the leaders of the urban agriculture movement recognize and are somewhat concerned about these attitudes, they are optimistic about the staying power of urban agriculture. To the question, "Will urban agriculture be able to maintain agroecological production when and if conventional agriculture makes an economic comeback?" the principal leaders of the urban agriculture movement, Adolfo Rodríguez Nodals and Nelso Companioni Concepción (2006), argue yes, provided that certain conditions are met:

1. The economic basis of urban agriculture is consolidated with efficient production and good incomes for the producers. For this to happen, direct sales to the public must continue to dominate and intermediaries must be eliminated or minimized. Their slogan is, "From the cantero to the consumer."

2. The system continues to recruit urban agriculturalists trained in and comfortable with the science and technology of agroecology.

3. The infrastructure and logistics of urban agriculture are strengthened and consolidated, in both input provision and output marketing.

Concerning infrastructure and logistics, the principal constraints relate to availability of adequate water, the production and transportation of organic fertilizers, and shortages of hard currency needed to purchase some inputs, such as irrigation equipment, in international markets. Whether these constraints can be overcome depends not only on political will and organization of the urban agricultural system, but also on natural and climatic factors and the Cuban government's ability to increase hard currency earnings through exports. The sustainable generation of adequate inputs will no doubt be a continuing challenge.

In any case, in light of its many beneficial effects on the environment, on community building, and on urban employment, urban agriculture will not be lightly discarded, even if conventional agriculture is revived in rural areas, as is likely to happen to some extent. In fact, some crops, such as potatoes, sugarcane, and tomatoes, lend themselves to industrial processing. Cuba has taken a pragmatic approach, continuing to grow these crops using conventional agricultural techniques to whatever extent the necessary inputs can be obtained.

The reactivation of agriculture in the countryside is on the government's agenda in order to increase Cuban food security, but this development is unlikely to negatively affect agriculture in the cities. As Rodríguez Nodals argued in a 2008 interview with CBS reporter Portia Siegelbaum, "I think that Cuba's urban agriculture is here to stay. That there is a little increase in the application of fertilizers and pesticides for specific crops is normal, but that is not to say that the country is going to shift away from organic farming to turn our organic gardens into non-organic ones" (Siegelbaum 2008).

That Cuba is serious about continuing along the path toward agroecological urban agriculture is evident from recent initiatives in this area. In his address to the Third Ordinary Session of the National Assembly on August 1, 2009, President Raúl Castro announced that a new program of suburban agriculture had been established and its first projects initiated in April 2009 in the capital of Camagüey province. Under the direction of the provincial government and with the participation of all relevant agencies, small farms are being established near the city but outside the boundaries of urban agriculture. The goal is to incorporate all lands within a reasonable distance of the city into food production, drawing on the city's population for the workforce while attempting to make frugal use of fossil fuels. This program was scheduled to expand to the municipalities of all provincial capitals starting in 2010. One of the principal institutional tools to be used in this expansion is Decreed Law 259, passed in July 2008, which provides for the distribution in usufruct of idle state lands to individuals and cooperatives. These lands are to be cleaned up of marabu and other past plants and put into production. Rapid development of suburban agriculture has taken place in 2010 in a process set to continue and expand to cover the entire territory—with the exception of a few completely urban or completely rural municipalities (Gayoso 2008; Hagelberg and Alvarez 2009; Koont 2010).

In introducing the new program to the National Assembly, Raúl Castro argued, "In this program, let us forget about using tractors and fuel, even if we had them in sufficient quantities. The idea is to carry out this program essentially with oxen—these are small farms—as a growing number of producers are already doing with excellent results. I have visited a few and can attest that they have converted their lands into true gardens where every inch of land is utilized" (Castro Ruz 2009, my translation). The program is conceived of as an extension to the suburbs of the principles underlying urban agriculture. In fact, in Castro's fairly long presentation to the assembly, Adolfo Rodríguez Nodals is the only Cuban official mentioned by name. Castro announced that Rodríguez and his small team at GNAU had been to lead the suburban agriculture effort due to the outstanding success of urban agriculture and the systematic and rigorous quarterly GNAU inspection visits that fostered this success. By August 4, in an interview about the results of GNAU's recently concluded forty-fifth inspection tour, Rodríguez was already being referred to as the head of the Urban and Suburban Agriculture Program (Varela Pérez 2009c).

In the Camagüey pilot project, 75 farms were ready to start operations in December 2009. By 2015 there are to be more than 1,400 small farms covering 52,000 ha—with roughly 750 of them raising cattle, 450 producing diversified crops, and 100 growing fruit- and timber-yielding trees as well as pigs, chickens, sheep, and goats (Febles Hernández 2009). In the meantime, the push is on to ready the teams of oxen and their drovers (*boyeros*), as a new chapter begins in the expanding influence of agroecological production for local consumption (Chávez 2009).

Is Urban Agriculture Spreading across the World?

At the time Cuba turned to urban agriculture, the practice was already widespread in some regions. In China, for example, growing food in the cities has a long historical tradition. Large populations and shortages of cultivable land have always meant that the densely settled cities had to fend for themselves food-wise to some extent. Although the Chinese Revolution of 1949 initially considered urban farms as a sign of backwardness and largely eliminated them, public authorities reversed that position beginning in the 1960s. By 1992, urban agriculture was feeding about one-third of the Chinese population, and most large Chinese cities were nearly self-sufficient in perishable foods. Urban-based food production was by then receiving one-third of state expenditures in agriculture, signaling significant government support for the movement (Mougeot 1994).

In most of the third world, urban agriculture has been and is still seen as part of a poverty-amelioration process, basically as something that contributes to the food security of poverty-stricken urban dwellers. The massive rural-urban migrations of the last half century have populated the cities of Africa, Latin America, and Asia with poor migrants from rural areas. These migrants know how to grow food, *and* are for the most part too poor to purchase even "cheap" industrially produced food. Naturally, they have turned to self-provisioning for survival. They have also participated in the urban food market by selling their surplus produce as peddlers or in more established farmers' markets, in order to generate the cash incomes absolutely necessary in an urban context. By now they have been joined by more strictly commercial, small-scale producers in the urban and peri-urban areas.

In response, international and national institutions focusing on food and development issues have started researching the urban agriculture phenomenon, funding projects and giving advice and support to local organizations involved in promoting this paradigm. Most prominent among these are the Food and Agriculture Organization of the United Nations (FAO); the UN Human Settlement Programme, a Nairobi-based UN agency established in 1978, also known as UN-HABITAT; and the

International Development Research Centre (IDRC) of Canada. The IDRC, the Canadian government's developmental-aid agency, has taken the lead in research and educational activities, funding projects, mostly in Africa and Latin America, that promote urban agriculture as a partial solution to the problem of sustainable urban living. It also has published books and articles reporting on and promoting urban agriculture.

In relatively land-rich Latin America, where land is not as scarce a resource as in Asia but is much more unequally distributed, urban agriculture has to date played a much smaller role. "Organic farming"—that is, cultivation without using chemical fertilizers or pesticides—has made some inroads, but mostly in rural settings and in export crops such as coffee, "organic-certified" vegetables, and tropical fruits. Mexico, for example, increased the area devoted to organic crops from 23,000 ha to 103,000 ha between 1996 and 2000. Organic crop-raising activities involved 33,000 producers in 262 different zones in 28 different states of Mexico, and generated U.S. $140 million in export earnings. This organic agriculture industry has developed with very little official support from the government. And fully 85% of the organic production in Mexico is exported to satisfy the growing demand for organically grown foods in niche markets in developed countries, primarily the United States. It is essentially unrelated to the domestic food production and distribution networks (Gómez Tovar et al. 2004).

To date, agroecological urban agriculture has made little headway in Latin America, although there have been efforts to promote it. For example, on April 10–20, 2000, IDRC, FAO, UN-HABITAT, UNDP, and other organizations cosponsored a seminar/workshop in Quito, Ecuador, entitled Urban Agriculture in the Cities of the Twenty-first Century. Participants, most of whom were mayors, came from 27 cities in Mexico, Honduras, Cuba, the Dominican Republic, Colombia, Ecuador, Brazil, Peru, Uruguay, and Argentina. The "Declaración de Quito" adopted by the seminar participants, including representatives of the GPAU of Ciudad de la Habana, recognized that although experience with urban agriculture in Latin America is still limited, it can make an important contribution to creating cities with equitable, healthy, and food-secure environments for their inhabitants. Among their recommendations were that cities recognize the important contributions of urban agriculture to strategies for social development; the generation of employment, incomes, and self-esteem; and environmental improvement and food security. In addition, urban agriculture should be taken into consideration as a component of soil use and environmental protection policies during urban planning. They stressed the need for support for urban agriculture from both local and national governments. They further proposed that technical training be provided to urban agriculturalists, and that adequate incentives be offered to interest urbanites in this line of activity (PGU-ALC Habitat Programa de las Naciones Unidas 2000).

The FAO Regional Office for Latin America and the Caribbean recently completed and published a study which concludes that urban agriculture could contribute to the availability and accessibility of high-quality foods in that part of the world but is hamstrung by the absence of three factors: supportive government policies and strategies, especially at the municipal level; institutional involvement; and participatory planning to ensure viability and sustainability in urban agriculture. Chile and Cuba are cited as the only two exceptions, and Chile only because of its success in introducing small-scale, commercial hydroponic units into the food economies of Santiago and Valparaiso as a micro-entrepreneurial activity. This success hardly constitutes a paradigmatic revolution in food production, but rather amounts to Chile having provided policy and institutional support that enabled some small-scale producers successfully to insert themselves into competitive commercial markets. This support came in training provided through institutions such as the Instituto de Desarrollo Agropecuario and the Centro de Educación y Tecnología de Chile, in market access for inputs and output, and in financial and commercial intermediation (Treminio 2004).

Is Industrial Agriculture Sustainable in the Long Term?

When Cuba decided—or had to decide—to turn to urban agriculture as the main source of fresh produce for urban consumption, it was truly blazing a new trail. Meanwhile, the rest of the world continued to move merrily along the path that Cuba had to give up. During the 1980s, the neoliberal, globalized, corporate-controlled economic order began to shape the policies of national governments and international institutions such as the World Bank, the International Monetary Fund, and the World Trade Organization, spreading industrialized agriculture across the globe.

Over the last decades, the industrial food production system has indeed lowered food prices, but as part of its legacy we must also count several less propitious outcomes (Wise 2010):

- Whether it is possible to continue these practices indefinitely is in serious doubt. They deplete nonrenewable resources, including in transportation, in a context of increasingly binding scarcities.
- It expels small rural farmers from their no longer competitive farms, generating marginalized populations unable to afford even the now-cheaper food.
- The neoliberal order, of which the global, multinational corporation–based industrial agriculture is a part, polarizes incomes in the world. This creates pressures to introduce new patterns of land use that cater to the demands of the rich, who keep getting richer.
- The agricultural technologies employed cause local environmental damage by degrading soils and water supplies, and cause global damage through the climatic effects of its heavy fossil fuel use and its role in diminishing biodiversity.
- It has toxic effects on workers during the production process, and on consumers via toxic pesticide residues in the marketed products.

As a result, agroecological, local production for local use has been gaining either forced or willing converts throughout the world—and is almost certain to gain more in the future as the consequences of industrial agriculture are absorbed and assimilated.

Resource Depletion

It takes on average 7–10 calories of fossil fuel energy to deliver 1 calorie of food to the U.S. dining table (Pollan 2007: 183). And the longer the producer-to-consumer chain, the worse this figure becomes. A 1 lb bag of prewashed California lettuce contains 80 calories of food energy. Drawing on the findings of Cornell ecologist David Pimentel, Michael Pollan (2007: 167) estimates that it takes 4,600 calories of fossil fuel to grow, harvest, chill, wash, package, and transport that pound of lettuce to a consumer on the U.S. east coast. This amounts to an astonishing 57:1 ratio of calories of fossil fuel used to food consumed.

Marginalization of Small Farmers

One factor driving the third-world mass migrations is the displacement of small farmers from lands appropriated for industrial monoculture cultivation of crops for the world market. These uprooted populations then migrate to urban areas in search of alternative livelihoods. As early as 1993, farmers began organizing to resist the corporate, globalized, neoliberal order that was wreaking havoc on small farming and indigenous communities all over the world. One of the most important organizations is a loose international association of local, small-scale peasants and farmers calling itself Vía Campesina. By 2008 it had grown to include more than 130 local organizations—including ANAP from Cuba—in more than 60 countries, and established an International Operative Secretariat in Jakarta, Indonesia (Hernández Navarro 2008). Although Vía Campesina is for the most part rurally based, its emphasis on local production of foodstuffs for local consumption; on agroecological, environmentally friendly production methods and technologies; and on promotion of biodiversity dovetails very well with urban agriculture.

Income Inequality and Loss of Food Security and Sovereignty

A full-blown food crisis befell the world in 2007–2008, when international prices for food staples such as rice, wheat, and maize skyrocketed. The World Bank reported in 2008 that global food prices had risen 83% over the past three years (Holt-Giménez et al. 2009: 6). On a trip to Cuba in July 2008, FAO General Director Jacques Diouf pointed out that food prices had risen 52% in less than a year. With regard to food-production inputs, there were cost increases of 78% for fertilizers, 60% for seeds, and 72% for animal feed (Guerra Rey 2008). According to Diouf, this crisis "was predictable and we predicted it, but it also was avoidable, and we could not avoid it" (Suárez Pérez 2008). Food riots occurred in Haiti, Indonesia, Mexico, Bangladesh, Burkina Faso, Egypt, Senegal, Cameroon, Morocco, Yemen, Somalia, and the Philippines (Urquhart 2008), reflecting the desperation and rage of people who were already earning less than $2.00 a day, not producing their own food, and spending 70% or more of their income to feed themselves (Altieri 2008), then were hit by a 50% or greater increase in food prices.

In their book *Food Rebellions!* Holt-Giménez, Patel, and Shattuck

(2009) list among the immediate causes of the crisis at least two that are related to fossil fuels or their attempted replacement by alternatives: the unprecedented jump—up to $145 per barrel—in the price of petroleum, and the increased production of agro-fuels—mainly ethanol—that either used or crowded out food crops. But the fundamental cause in their view is the neoliberal era's globalized, corporate-profit-driven reorientation of the world's food production system toward large-scale, monoculture-based industrial agriculture. Under the pressure of structural adjustment programs imposed by the World Bank and the International Monetary Fund, free trade agreements, and other such policies, most third-world countries have lost their ability to control their food production locally. Instead, they have become subject to the vagaries of the international marketplace. So, for example, it was perhaps foreseeable that corn would be cultivated to produce fuel for cars in rich lands rather than food for the insolvent poor in the third world.

The full-blown crisis has strengthened the positions of organizations like Vía Campesina that argue for major changes in the world system of food production in order to move away from fossil fuel dependency and eliminate the control of profit-driven agribusinesses over the production and distribution of food crops (Altieri 2008; Vivas 2008; Weis 2008). Perhaps the most important concept that Vía Campesina formulated and introduced into public discourse was *food sovereignty:* the right of each people to define their own policies concerning agriculture, to protect and regulate their national agricultural production and markets with the aim of sustainable development, to decide to what extent they want to be self-sufficient in food, and to prevent their domestic markets from being inundated with subsidized products from other countries. The emphasis is on local, ecologically sustainable production of culturally appropriate, wholesome, and nutritive foods. Thus conceived, food sovereignty incorporates the concept of food security (adequate food supplies to meet the population's needs) and even overlaps with national security. Reliance on international markets dominated by powerful countries and huge corporations raises fairly obvious economic and political concerns, ranging from disruption of local markets by manipulated world prices to the potential use of food supplies as instruments of political manipulation or pressure. For these reasons Bolivia, Venezuela, and Nepal have inserted the concept of food sovereignty into their national constitutions (Hernández Navarro 2008). Even at intergovernmental levels, there is growing recognition that food security and food sovereignty concerns have to be considered in global food production and trading systems (Bradsher 2008; Concepción Pérez 2008; Khor 2008; Spieldoch 2008).

In the first world, the food security aspect so crucial in the third world is a secondary concern, because per capita incomes are in general high enough to meet basic needs. It is still a concern, however. In the United States, for example, a large portion of those living in poverty are maintained in the industrial food distribution networks through the food stamps component of welfare. About one-seventh of the U.S. population, or almost 43 million people, participated in this program at the end of 2010 (Smith 2010). However, it is estimated that about one-third of eligible people do not participate, and thus do not receive any food stamps (FRAC 2008), and this does not include the sizable population of ineligible persons. This group, composed of undocumented immigrants, was estimated to be around 10 million in 2010 (StateMaster.com 2010). And undocumented immigrants are more likely than the general population to live in extreme poverty. This sector has the same set of interests and goals in their struggle for food security as the unwelcome rural migrants in the cities of the third world.

Food Quality Concerns

The other, much larger sector in the developed world, the middle and upper classes who do not suffer from food insecurity, are also showing signs of beginning to question both the long-term viability and the current adequacy of the existing food supply networks of industrial agriculture. As a visit to the organic produce section of any supermarket or to a farmers' market selling local produce will confirm, concern over the sustainability, ecological footprint, and personal health implications of industrially processed foods has created niche markets for organically produced foods for consumers who can afford and are willing to pay a premium for these products. In addition, some cooperative- and community-garden-based, local growing of food has also started taking place in Europe and the United States.

Ecological Concerns

One fact serves to drive home the global ecological consequences of modern agriculture: it has been estimated that industrial food production, processing, and distribution (if the effects of associated deforestation are included) accounts for about 50% of all greenhouse gases emitted due to human activity, a major environmental concern worldwide (Vivas 2009). The changes necessary to lessen dependence on fossil fuel go well beyond the issue of organic production. It is not just a matter of substituting organic alternatives for petroleum-based pesticides and fertilizers. For in the current agricultural system, much of the fossil fuel energy is consumed by large-scale, labor-saving machinery, which is only usable in monoculture cultivation, and by the associated requirement to transport foods under refrigerated conditions to distant retail markets. Shortening the chain that connects crop production to consumer does more than reduce fossil fuel consumption. By localizing and reducing the scale of production, it also eliminates intermediaries in distribution and lessens the role of agribusinesses. It encourages and supports ecologically sustainable, local farming practices and small farmers.

Since 1990, a struggle has been developing between two distinct visions of the future of farming and food systems: globalized agribusiness versus localized small farming. The recent food crisis has shifted the balance in this struggle toward the latter option. That option is increasingly seen not as a nostalgic attempt to turn back the clock to a simpler time but rather as a scientifically based, ecologically sound path toward an equitable, sustainable food future for humankind. Recent research has increasingly confirmed that agroecological and local production of foodstuffs for local use is a viable option for our planet's human population (FAO 2007; Halweil 2006). In any case, sooner or later the fossil fuels will run out. And, in the meantime, their heavy use has many well-known environmentally deleterious effects. As a result, it can be argued that Cuba's urban agriculture is no longer an aberration or an outlier. Instead, the model of small-scale agriculture whose consumers live next door is becoming more a successful example worthy of emulation than a desperate attempt at survival that worked.

What Can the World Learn from Cuban Urban Agriculture?

There can be little doubt that Cuba's experience with urban agriculture in the Special Period makes for an interesting case study of an impressive effort. In fact, this effort is really not just impressive, but unique and exemplary. Alone among the myriad urban agriculture processes throughout the world, it is almost completely agroecological, and it uses fossil fuels minimally, even in transportation. It is based on small-scale, locally sourced, and locally consumed production—in other words, it is the complete antithesis of industrial agriculture as practiced across the world. It is exemplary because of its comprehensive organization as a food production system for the entire urban population of the nation. Facing desperate food shortages, Cuban society and the country's political and governmental institutions responded with an urban agricultural system capable of making significant contributions to the food security and food sovereignty of the Cuban nation. As argued in chapters 3 to 6, this success was enabled by a number of factors: the general educational preparedness level of the Cuban population; the capacities in research and development and in basic scientific investigation built up over decades; the solid network of governmental institutions that allowed the rapid building of an organizational scaffold for urban agriculture; widespread training in the new technologies through extension services; the organization of a network for input provision to the decentralized system of production units; and, last but not least, moral and material incentive structures intended to generate enthusiasm and effort among the dispersed producers.

Actually, the urban physical location of this agricultural revolution turns out to be of secondary importance. What is more important about Cuban urban agriculture is that it represents a paradigm shift away from the fossil-fuel and capital-intensive technologies of industrial agriculture. Cuba did not voluntarily embrace this shift but was forced into it by the scarcities of the Special Period. Agriculture had to be organized along more labor-intensive lines in order to replace petroleum and its derivatives in the production and distribution phases of agriculture. Urban locations were a good fit for all the requirements: there was a large, available labor force; the setting demanded nontoxic production technologies due to the proximity of dense human populations; and no transportation of the output over large distances was necessary. Cuban urban agriculture is located in or near population centers, but its first and most important hallmark is its completely agroecological nature and its minimal use of fossil fuel resources in the distribution of its product. Another hallmark, perhaps equally characteristic of Cuban urban agriculture, is the presence of a strong and stable central organizational structure, derived from the national government's strong commitment to this new paradigm and its success. This strong commitment exhibited itself in the rapid redirection of resources in research and development, in science and technology, and in higher education to support the transition. New institutions, policies, and practices created a dynamic new environment facilitating the rapid development of urban agricultural production in an eco-friendly manner.

Could the example of Cuban urban agriculture be replicated in other countries? Probably not in the short run and not completely. Most other countries exhibit neither the critical urgency and need nor the full readiness and capacity that made this transition both unavoidable and possible in Cuba. For one thing, the agroecological paradigm entails employing a much larger fraction of the labor force in the agricultural sector than is true in industrial agriculture. In most countries practicing advanced industrial agriculture, only 1%–2% of the workforce is employed in agriculture. Cuba already employs 7% of its workforce in urban agriculture alone—not counting the home gardeners who grow their own food. Inherent in more labor-intensive agriculture is higher food costs, especially if the workers are to receive a living wage.

This said, however, the Cuban example of the last two decades provides many features of great relevance and importance for peoples who are struggling for food security and sovereignty, particularly in the third world. First among the lessons of Cuba is that national governments must be committed to producing a less hostile, even friendly, and more supportive environment for urban agriculture. This would imply a shift in priorities that guide allocation of the (as a rule rather meager) government resources in such areas as research and development, science and general education, technological innovation, and training relevant to urban agriculture. Second, the transition to agroecological production is possible and must be encouraged in order to overcome the limitations on urban production currently posed by toxic chemical inputs. Finally, a shift in attitudes is needed to overcome societal and governmental resistance to incorporating food production into the permissible set of urban activities.

For the rich, first-world countries contemplating such a paradigm shift, perhaps the most important lesson is that an agricultural system that employs large numbers of well-paid urban producers using labor-intensive, agroecological technologies is, in fact, possible. In such a paradigm, consumers will spend a larger fraction of their income on food but will receive new individual benefits in the form of wholesome, high-quality food, and collective benefits in the form of preservation and protection of the environment and biodiversity. Developed countries already have sizable government resources at their command, with taxation rates, as a percentage of GDP, that range from the mid-30s to above 60 percent. They already have well-educated populations and considerable existing physical and social infrastructure. They also have formidable capacities in research and development, technological innovation, and higher education. They would only need to redirect these efforts to those areas that favor the desired changes in the agricultural paradigm, as Cuba did.[2]

Still, the resistance such a shift will encounter is formidable. It will require many changes in institutions and in laws and their enforcement—for example, modification of zoning laws controlling the use of land for growing food or imposition of new regulations to prevent the potential for mutual pollution of urban agriculture by other urban activities and vice versa. If adopted, such a paradigm would also trigger major economic changes in, for example, the sources and levels of incomes earned by different groups of economic actors, and in the way these incomes are spent.

In the face of these challenging obstacles, which may even be impossible to overcome given the existing relations of power and influence, developed countries can at least see a working, real-life example in Cuba of how an urban, agroecological food production system functions with at least partial success. And even if the obstacles cannot be removed all at once, at least some initial steps can be taken fairly easily, such as widespread composting for use in household gardens and the practice of intercropping and biodiversity in such gardens. Such initial steps may promote a more favorable public attitude toward the prospect of a paradigm shift.

In the meantime, it behooves all of us to continue watching closely the evolution of Cuba's at once necessary *and* principled experiment in agroecological, local production of foodstuffs for local use, as it is bound to generate new lessons for the world with its creative innovations and responses to the challenges it continues to face. Whether or not they all are completely successful, these responses represent the collective reaction of the Cuban people to adverse circumstances beyond the country's control, and they are and will continue to be proof of its resilience.

Notes

Chapter 1. Cuban Agriculture

1. The head of INRE, General Moisés Sio Wong, reminded Raúl Castro of this story when Castro made a return visit ten years later.

2. The three merged organizations were the Instituto de Agronomía (Institute of Agronomy), the Instituto de Mejoramiento de Plantas (Institute for the Improvement of Plants), and the Instituto de Investigaciones Tropicales (Institute for Research on the Tropics)—all established in 1966. The first two institutions were outgrowths of the old Estación Central Experimental Agronómica (Central Experimental Station for Agronomy) located in Santiago de las Vegas, initially founded in 1904, then revitalized in 1959 with the triumph of the revolution (Díaz Otero and García Capote 2006: 11).

Chapter 2. The Nature and Organization of Cuban Urban Agriculture

1. Interview with Nelso Companioni Concepción by the author at GNAU headquarters, March 26, 2007.

2. CSS Fortalecidas (CSSFs) have more formal structure and organization (such as strict accounting practices) than regular CSSs. Any CSS can become fortalecida by meeting set requirements.

3. By the end of 1990s, organopónicos were attracting international attention (see, e.g., Hoffmann 1999).

4. The contributing institutions are INIFAT, the Instituto de Investigaciones en Sanidad Vegetal (INISAV, Institute for Plant Health Research), the Instituto de Investigaciones de Riego y Drenaje (IIRD, Institute of Research on Irrigation and Drainage), the Instituto de Investigaciones Hortícolas "Liliana Dimitrova" (IIHLD, Liliana Dimitrova Institute for Horticultural Research), the Instituto Nacional de Ciencias Agrícolas (INCA, National Institute of Agricultural Sciences), and the Instituto de Suelos (IS, Institute of Soils).

5. The guidelines were revised and approved during a 2004 meeting of GNAU attended by Juan Pérez Lama, vice minister of the Ministry of Agriculture, and Yadira García, a member of the Political Bureau of the Communist Party of Cuba (GNAU 2007a: 88).

6. Culantro is a biennial culinary herb, *Eryngium foetidum*, similar to cilantro and widely used in Caribbean and Latin American cooking.

7. CDRs are essentially neighborhood organizations that try to mobilize local residents in support of revolutionary goals and for social purposes. In urban agriculture they play an important role in promoting backyard production.

8. For example, the province of La Habana has four active state agricultural enterprises: Hortícola Metropolitana (Metropolitan Horticulture), Cultivos Varios Habana (Varied Crops Havana), Pecuaria Bacuranao (Bacuranao Livestock), and Aseguramientos y Servicios de la Agricultura Urbana (Supplies and Services for Urban Agriculture) (Acosta Mirrelles 2006a: slide 19). Each granja urbana in the 15 municipios of the province is affiliated with one of these four enterprises.

Chapter 3. Foundations in Education, Research, and Development

1. This is a technique used in the asexual reproduction of plants via cloning of tissue taken from a single plant.

2. Acarology is the study of mites and ticks; nematology is the study of roundworms.

3. Traditional banana plant spacing is 7–8 m^2.

4. HeberNem acts on the eggs and larvae of nematodes in two ways: via the enzyme chitinase produced in the medium where the bacteria are grown, and via hydrogen sulfide produced by the bacteria themselves (Mena Campos et al. 2007: 52). Chitinase damages the outer layers of the eggs and larvae, and hydrogen sulfide induces gaseous vacuoles in the eggs and larvae, impeding their development into surviving adults.

Chapter 4. Restructuring Worker Training, Preparatory Education, and Material Inputs for Urban Agriculture

1. Interview by the author with Eugenio Fuster Chepe, president of ACTAF, at ACTAF headquarters, April 3, 2006.

2. The two tasks that do not directly mention training or extension are the strengthening of local ACTAF affiliates and the establishment of a culture of consumption of fresh vegetables and dried herbs, as well as of preservation of vegetables (ACTAF 2005).

3. For instance, it was precisely to these three CPAs that INCA brought its participatory plant-breeding program in Havana province (Taset Aguilar 2005).

4. The exceptions are the municipalities of Havana city, which are to have no farms in the central municipalities and only one in each of the peripheral municipalities. A few other municipalities, such as Varadero and Ciénaga de Zapata in Matanzas, are also exempted.

Chapter 5. Creating Material and Moral Incentives to Motivate Workers

1. This revision is Resolution 9, issued by the Ministry of Work and Social Security, and titled "Reglamento general sobre las formas y sistemas de pago" (General regulation on the forms and systems of pay).

2. There are some legal restrictions on goods that can be sold in MALs. Completely excluded, for example, are beef, milk products, and coffee.

Chapter 6. Technological Innovation in Urban Agriculture

1. The technical definition is as follows: suppose n associated crops are grown intercropped on a given plot of size A, measured, say, in hectares. The plot yields output quantities Q1, Q2, . . . , Qn measured, say, in tons of the n associated crops. Each crop has a fixed land productivity ratio LPC|sub/sub|, measured, say, in tons per hectare, in monoculture cultivation. Then the area required to produce the quantity Qi in monoculture can be calculated as mi = Qi/LPCi [tons/(tons/hectare) = hectares], and M = m1 + m2 + . . . + mn represents the total area needed in monoculture to produce the combination of outputs Q1, Q2, . . . , Qn. The ratio of this total area M to the area A required in intercropping is called the land-equivalent ratio LER, a pure number that does not depend on the units chosen to measure area or output. If LER >1, more area is required to produce the identical output in monocultures compared with intercropping, and the latter is more efficient than the former.

Chapter 7. Case Studies of Urban Agriculture

1. Maternity homes (*hogares maternos*) are homes for women with high-risk pregnancies or birth complications, where women receive better food and constant medical supervision. Gastronomía Familiar is a program that provides lunches to elderly people in their homes, while workers receive subsidized lunches at their workplaces.

2. Magnetization functions as a water conditioner that has been shown to increase the growth and health of both livestock and crops.

3. Everyone I met always referred to him simply as "Falcón." I never did learn his full name.

Chapter 8. Evaluating the Success of Cuban Urban Agriculture

1. Personal interview with Nelso Companioni Concepción by the author at GNAU headquarters, March 26, 2007.

2. The Cuban population increased little over the decade, so new construction or formation of net new households is not a significant factor.

3. Given that Havana City province has about 20% of Cuba's population but only about 3% of urban agricultural lands, it is a foregone conclusion that it will have greater difficulties in meeting the food needs of its population.

4. These are my calculations based on data in the *Lineamientos* (GNAU 2007a).

5. In 2005, Cubans consumed 3,356 calories, 88 g of protein, and 45 g of fats daily. Non-market sources—namely, the ration book, public (including lunches at workplaces) and social consumption, and self-provisioning—accounted for 3,063 calories (90% of the total), 78 g of protein (72% of the total), and 32 g of fat (72% of the total). Neither state MAEs nor free MALs are involved in these distributional channels.

6. For a case study in the Cerro municipality of Havana, see Smit and Bailkey 1996.

Chapter 9. Looking to the Future of Urban and Sustainable Agriculture

1. Recall that San José de las Lajas is the site of the Agrarian University of Havana and three national, government-sponsored agricultural research centers: ICA, CENSA, and

INCA. Thus, San José de las Lajas is one of the centers of dissemination of the science of urban agriculture.

2. This might not be such an outlandish idea. The experience of small-scale production in World War II victory gardens may still linger in the historical memories of the U.S., British, and German populations. It is estimated that about 40% of fresh produce in the United States during the war years was grown in household gardens.

Bibliography

Academia de Ciencias de Cuba (ACC)

2009   "Academia de Ciencias de Cuba." http://www.academiaciencias.cu/ (accessed March 30, 2009).

Acosta, José A., Ernesto L. Mola, Marianela G. Siverio, Carlos B. Nordelo, and Luis H. Martínez

2007   "Novel technology for sustainable agriculture." Camagüey: CIGB. http://www.gndp.cigb.edu.cu/Press%20Release/nematicide.htm (accessed May 29, 2009).

Acosta Mirrelles, Orlando

2006a   "La agricultura urbana en los municipios de Ciudad de la Habana con enfoque sostenible y agroecológico." PowerPoint presentation. http://www.cipotato.org/urbanharvest/documents/pdf/La%20Habana%20-&20Orlando%20Acosta.pdf (accessed July 8, 2009).

2006b   "Misión al 2007." *Agricultura Orgánica* 12(2): 14–15.

Agencia de Información Nacional (AIN), Cuba

2001   "La Habana, provincia de mayor puntuación en la agricultura urbana." *El Habanero* (Havana), October 10. http://www.elhabanero.cubaweb.cu/2001/octubre/nro222_01Oct/inf_1oct292.html (accessed June 9, 2009).

2008   "Cuba: Ratifica sus ventajas del cultivo protegido." *Granma*, February 2. http://www.freshplaza.es/news_detail.asp?id=3606 (accessed August 4, 2008).

Ahmed, Belal, and Sultana Afroza

1996   *The political economy of food and agriculture in the Caribbean.* Kingston, Jamaica: Ian Randle.

Altieri, Miguel A.

1995   *Agroecology: The science of sustainable agriculture.* Boulder, Colo.: Westview Press.

2000   "Agroecology: Principles and strategies for designing sustainable farming systems." http://www.agroeco.org/doc/new_docs/Agroeco_principles.pdf (accessed February 16, 2010).

2008   "Movilizándonos para rescatar nuestro sistema alimentario." http://alainet.org/active/23532 (accessed March 23, 2010).

Altieri, Miguel A., Nelso Companioni, K. Cañizares, C. Murphy, P. Rosset, M. Bourque, and C. I. Nicholls

1999   "The greening of the 'barrios': Urban agriculture for food security in Cuba." *Agriculture and Human Values* 16(2): 131–40.

Alvarado de la Fuente, Fernando

2008 "Invertir en ecológicos o invertir en transgénicos?" *Rebelión*, August 30. http://www.rebelion.org/noticia.php?id=71927 (accessed August 30, 2008).

Alvarez, José

2004a *Cuba's agricultural sector*. Gainesville: University Press of Florida.

2004b "The issue of food security in Cuba." Gainesville: Florida Cooperative Extension Service, Department of Food and Resource Economics.

2009 "Environmental deterioration and conservation in Cuban agriculture." Institute of Food and Agricultural Sciences Extension Publication No. FE489. University of Florida Electronic Data Information Source. http://edis.ifas.ufl.edu/fe489 (accessed August 5, 2010).

Anderson, Pamela K., and Francisco J. Morales, eds.

2005 *Whitefly and whitefly-borne viruses in the tropics: Building a knowledge base for global action*. Cali, Colombia: Centro Internacional de Agricultura Tropical.

Arce Bejarano, Madalys

2009 "Se consolida en Minas la agricultura urbana." *Radio Minas Digital*, June 4. http://www.radiominas.icrt.cu/index.php?option=com_content&task=view&id=873&Itemid=1 (accessed June 9, 2009).

Armario Aragón, Danneys, Bladimir Díaz Martín, Alianny Rodríguez Urrutia, Joaquín Machado de Armas, J. M. Portieles Rodríguez, Alberto Espinosa Cuéllar, Osvaldo Triana Martines, and Juan R. Gálvez Guerra

n.d. "El humus: Una alternativa de fertilización en la producción de bananos plantados en altas densidades." Instituto de Investigaciones en Viandas Tropicales (INIVIT). http://74.125.47.132/search?q=cache:ukEVjQ2nsf8J:www.villaclara.cu/UserFiles/File/Portal%2520prov./CIENCIA/Reporte2.doc+%E2%80%9CEl+Humus:+Una+alternativa+de+fertilizaci%C3%B3n+en+la+producci%C3%B3n+de+bananos+plantados+en+altas+densidades.%E2%80%9D&cd=1&hl=en&ct=clnk&gl=us (accessed July 2, 2009).

Armario Aragón, Danneys, Sergio Rodríguez Morales, Sinesio Torres García, Alberto Espinosa Cuellar, Osvaldo Triana Martines, Lourdes Cabrera Tamayo, and Pablo R. Lago Gato

n.d. "Una alternativa aplicable a sistemas extradensos de bananos." Instituto de Investigaciones en Viandas Tropicales (INIVIT). http://www.villaclara.cu/UserFiles/File/Portal%20prov./CIENCIA/bananos.doc (accessed March 23, 2009).

Armengol, Alejandro

2008 "Aumentan áreas de agricultura urbana." *Cuaderno de Cuba*, May 15. http://www.cuadernodecuba.com/2008/05/aumentan-reas-de-agricultura-urbana.html (accessed October 17, 2009).

Arteaga, Carlos

2007 "Memorias del taller nacional de reordenamiento de las unidades básicas de producción cooperativa (UBPC) y su transformación en fincas agroecológicas." *Agricultura Orgánica* 13(2): n.p.

Asociación Cubana de Técnicos Agrícolas y Forestales (ACTAF)

2004 "Estatutos: II Congreso." October 2004. Havana.

2005   "Breve apuntes sobre el acompañamiento de ACTAF en el recorrido del Grupo Nacional de Agricultura Urbana." *Agricultura Orgánica* 11(1): n.p.

2006a   "Funcionamiento." *Agricultura Orgánica* 12(2): 24–25.

2006b   "Mi Programa Verde." Paper presented at the sixth Encuentro de Agricultura Orgánica y Sostenible. Compact disk, ISBN: 959-0282-17-5. Havana.

2009   "Qué es la ACTAF?" http://www.villaclara.cu/paginas-amarillas/actaf (accessed June 2, 2009).

Asociación Cubana de Técnicos Agrícolas y Forestales (ACTAF), Filial Provincial Villa Clara

2006   "Tarea: Maestros viajeros de la ACTAF." http://ebookbrowse.com/ tarea-maestros-viajeros-doc-d43945187 (accessed June 2, 2009).

Asociación Nacional de Agricultores Pequeños (ANAP)

1999   "ANAP." *Agricultura Orgánica* 5(3): 32–33.

2006   "Campesino a campesino." Plenary presentation at the II Encuentro Nacional de Agroecología. http://www.campesinocubano.anap.cu/2006/noviembre/ 29informe.htm (accessed June 2, 2009).

2009a   "Centro Nacional de Capacitación de la ANAP: 'Niceto Pérez García,'" http:// www.campesinocubano.anap.cu/secciones/escuela.htm (accessed June 2, 2009).

2009b   "Estructura." http://www.campesinocubano.anap.cu/secciones/estructura.htm (accessed June 15, 2009).

Asociación Nacional de Innovadores y Racionalizadores (ANIR)

2009   "Reglamentos." ANIR Web site. http://www.redciencia.cu/anir/Reglamento%20 ANIR.pdf (accessed March 28, 2009).

Avendaño, Bárbara

2006   "Biofábricas: Miniaturas gigantes." *Bohemia,* October 5. http://www.bohemia. cu/2006/10/05/cienciatecnologia/biotecnologia.html (accessed May 29, 2009).

2007   "No son un adorno intelectual." *Bohemia,* October 17. http://www.bohemia. cu/2007/10/15/cienciatecnologia/ciencias-sociales.html (accessed April 23, 2009).

Bakker N., M. Dubbeling, S. Gundel, U. Sabel-Koschella, and H. de Zeeuw, eds.

2000   *Growing cities, growing food: Urban agriculture on the policy agenda.* Feldafing, Germany: German Foundation for International Development (DSE).

Baños Fernández, Alipio Heriberto

2009   "Resultados positivos en la visita del Grupo Nacional de la Agricultura Urbana." *Ecos de Mantua* 65, March 15. http://www.ecosdemantua.cu/titulares. php?Fecha=15/03/2009&ID=610&l=s (accessed June 9, 2009).

Barraclough, Solon L.

2000   *Meanings of sustainable agriculture: Some issues for the south.* Geneva: South Centre.

Barreras Ferrán, Ramón

2008   "Celebran décimo aniversario del movimiento de la agricultura urbana." *5 Septiembre,* January 2. http://www.5septiembre.cu/agricultura127.htm (accessed July 23, 2008).

Barrio, Margarita, Dora Pérez, Yahili Hernández, Luis Raúl Vázquez, Osviel Castro, and Isis Sánchez
2009   "Los amores que le faltan a la tierra." *Juventud Rebelde,* February 15. http://www.juventudrebelde.cu/cuba/2009-02-15/los-amores-que-le-faltan-a-la-tierra-i/ (accessed February 15, 2009).

Barthelemy, Silvia, and Marnie Fiallo Gómez
2006   "Impactos en ingeniería genética y biotecnología." *Atina Arica,* May 27. http://www.atinaarica.cl/content/view/957/Impactos-en-Ingenieria-Genetica-y-Biotecnologia.html (accessed May 26, 2009).

Batista, Pastor
2009   "Incremento del área semiprotegida." *Granma,* June 8. http://www.granma.cubaweb.cu/2009/06/08/nacional/artic04.html (accessed June 8, 2009).

Bell Lara, José
2002   *Globalization and the Cuban revolution.* Trans. Richard A. Dello Buono. La Habana: Instituto Cubano del Libro.

Borrego, Juan Antonio
1999   "Arroceros de Sur del Jíbaro comenzaron siembra de frío." *Granma,* November 20. http://www.granma.cubaweb.cu/20nov99/nacional/articulo1.html (accessed November 22, 1999).

Borroto, Carlos, Gil Enríquez, and Merardo Pujol
2001   "Seguridad alimentaria, semillas y biotecnología: El caso de Cuba." Centro de Ingeniería Genética y Biotecnología (La Habana). http://www.redbio.org/portal/encuentros/enc_2001/mesaredonda/M-03/Merardo%20Pujol/M-03.pdf (accessed July 3, 2009).

Bosch, Hernán
2006   "Agroecología: Más que una opción, una estrategia." Asociación Nacional de Agricultores Pequeños (ANAP). http://www.campesinocubano.anap.cu/2006/noviembre/estrategia.htm (accessed June 2, 2009).

Bradsher, Keith
2005   "Breves apuntes sobre el acompañamiento de ACTAF en el recorrido del Grupo de Agricultura Urbana." *Agricultura Orgánica* 11(1): n.p.
2008   "Trade talks broke down over Chinese shift on food." *New York Times,* July 31.

Brigadas Técnicas Juveniles (BTJ)
2009   "Reglamento: Brigadas Técnicas Juveniles." Website of Brigadas Técnicas Juveniles en Camagüey. http://www.cmg.jovenclub.cu/btjpag/regla.php (accessed July 6, 2009).

Brizuela Roque, Ramón
2008   "La ciencia, una 'caja de milagros.'" *Matahambre.* Weblog de Ramón Brizuela Roque, July 24. http://matahambre.blogia.com/2008/072401-la-ciencia-una-caja-de-milagros-.php (accessed April 25, 2009).
2009   "Apoyan científicos recuperación agrícola." *Trabajadores,* April 25. http://www.trabajadores.cu/materiales_especiales/coberturas/gustav-epopeya-de-la-recuperacion/apoyan-cientificos-recuperacion-agricola (accessed July 8, 2009).

Buscagro.com

2008 "La biotecnología cubana ofrece el nematicida biológico HeberNem." November 27. http://www.buscagro.com/blog/528-la-biotecnologia-cubana-ofrece-el-nematicida-biologico-hebernem/ (accessed February 25, 2009).

Cabrera Balbi, Pausides

2009 "Debaten deficiencias y reformas para agromercados." *Granma Digital*, October 5. http://www.cuba.cu/noticia.php?archivo_noticia&id=3191.

Cadena Agramonte

2007 "Cuba: Prevén incremento de hortalizas en la provincia de Granma." September 18. http://www.freshplaza.es/news_detail.asp?id=333 (accessed March 23, 2010).

Carrión Fernández, Miriam

2006 "General y agricultor, una experiencia: Entrevista con el General de Brigada Moisés Sio Wong." *Agricultura Orgánica* 12(2): 2–3.

Casanova, Antonio, Adrián Hernández, and Pedro L. Quintero

2002 "Intercropping in Cuba." In *Sustainable agriculture and resistance: Transforming food production in Cuba*, ed. Fernando Funes Aguilar, Luis García, Martin Bourque, Nilda Pérez, and Peter Rosset, 144–53. Oakland, Calif.: Food First Books.

Castro Ruz, Fidel

1960 "Discurso pronunciado por el comandante Fidel Castro Ruz, primer ministro del gobierno revolucionario, en el acto celebrado por la Sociedad Espeleologica de Cuba, en la Academia de Ciencias, el 15 de enero de 1960." *Granma Digital*, http://www.cuba.cu/gobierno/discursos/1960/esp/f150160e.html (accessed March 18, 2011).

Castro Ruz, Fidel, and Ignacio Ramonet

2008 *My life: A spoken autobiography*. New York: Scribner.

Castro Ruz, Raúl

1997 *Desatar los nudos que atan el desarrollo a las fuerzas productivas*. Havana: ACTAF.

2009 "Tiene que ser el pueblo, con su Partido a la vanguardia, el que decida." Address to the National Assembly of Popular Power, August 1, 2009. *Granma*, August 1. http://www.granma.cubaweb.cu/2009/08/01/nacional/artic19.html. (accessed August 2, 2009.

Centro de Estudios para la Transformación Agraria Sostenible (CETAS)

2008 "Cinco años del Centro de Estudios para la Transformación Agraria Sostenible: Universidad de Cienfuegos." http://cetas.ucf.edu.cu/perfil_general.htm (accessed March 28, 2009).

Centro de Ingeniería Genética y Biotecnología (CIGB)

2007 "Profile." http://gndp.cigb.edu.cu/GNDP%20last/profile.htm (accessed May 26, 2009).

2008 "Development of an effective biological nematicide, HeberNem, based on *Tsukamurella paurometabola* C924 strain." http://gndp.cigb.edu.cu/portfolio2008/Development%20of%20an%20effective%20biological%20nematicide,%20

HeberNem,%20based%20on%20Tsukamurella%20p%20aurometabola%20
C924%20strain.htm (accessed May 29, 2009).

Centro de Ingeniería Genética y Biotecnología (CIGB) de Camagüey

2007   "Manual de aplicación del bionematicida HeberNem." http://revistas.mes.edu.
cu/eduniv/events/fitosanidad/Manual-Aplicacion-Bionematicida-HeberNem.
pdf/at_download/file (accessed May 19, 2009).

Centro de Ingeniería Genética y Biotecnología (CIGB) de Sancti Spíritus

2007   "Nuestro Centro." http://ss.cigb.edu.cu/Inicio.htm (accessed May 27, 2009).

Centro de Investigaciones de la Economía Mundial

2004   *Investigación sobre ciencia, tecnología y desarrollo humano en Cuba 2003*. Havana: Centro de Investigaciones de la Economía Mundial, Programa de Naciones Unidas para el Desarrollo.

Chávez, Armando Sáez

2009   "Con los bueyes, pero halando parejo." *Granma*, November 4. http://www.granma.
cubaweb.cu/2009/11/04/nacional/artic01.html (accessed November 4, 2009).

Choy, Armando, Gustavo Chui, and Moisés Sio Wong

2005   *Our history is still being written: The story of three Chinese-Cuban generals in the
Cuban revolution*. New York: Pathfinder Press.

Codorniú Pujals, Daniel

1998   "Ciencia e innovación tecnológica en Cuba: Estado actual y proyecciones." IN-
NRED. http://www.innred.net/iber/Eventos/1998/C98_02.htm (accessed April
23, 2009).

2001   "Como lo soñó Fidel." *Granma*, January 2. http://granma.cubasi.cu/temas10/
articulo5.html (accessed April 23, 2009).

Coloane, Juan Francisco

2008   "Las claves de un sistema que se despedaza." *Rebelión*, June 6. http://www.
rebelion.org/noticia.php?id=68490 (accessed June 6, 2008).

Companioni Concepción, Nelso

2006   "Particularidades del movimiento extensionista en la agricultura urbana." *Agricultura Orgánica* 12(2): 30–32.

2007   "La agricultura urbana: Un sistema alternativo de producción de alimentos en
Cuba." PowerPoint Presentation at INIFAT, Havana.

Companioni Concepción, Nelso, Yanet Ojeda Hernández, Egidio Páez, and Catherine
Murphy

2002   "The growth of urban agriculture." In *Sustainable agriculture and resistance:
transforming food production in Cuba*, ed. Fernando Funes Aguilar, Luis García,
Martin Bourque, Nilda Pérez, and Peter Rosset, 220–36. Oakland, Calif.: Food
First Books.

Concepción Pérez, Elson

2008   "Llamado urgente para acciones que no pueden demorar." *Granma*, June 6.

Corbière, Emilio J.

2008   "Pronostican una escalada de costos de la producción agrícola." *Argenpress*, July
24. http://www.argenpress.info/notaprint.asp?num=057316&parte=0 (accessed
July 24, 2008).

Cornide Hernández, María Teresa, and Rodobaldo Ortiz Pérez

2006   "Desarrollo de los estudios genéticos en plantas como base científica de la actividad de fitomejoramiento." In *Las investigaciones agropecuarias en Cuba: Cien años después,* coord. María Teresa Cornide Hernández, 175–206. Havana: Editorial Científico-Técnica.

Cruz, María Caridad, and Roberto Sánchez Medina

2003   *Agriculture in the city: A key to sustainability in Havana, Cuba.* Ottawa: IDRC/ Ian Randle.

Cuba, Adela

2002   "Participación en la esfera de sanidad vegetal en el movimiento cooperativo de la agricultura urbana." Master's thesis, FLACSO, Universidad de Havana.

Cuban Daily News

2008a   "In Cuba the efficient production of food is a task of prime importance." June 22. http://www.cubaheadlines.com/2008/06/22/12010/in_cuba_efficient_production_food_a_task_prime_importance.html (accessed December 2, 2008).

2008b   "In the Cuban provinces urban ag surpasses vegetable crop." August 6. http:// www.cubaheadlines.com/2008/08/06/12775/in_cuban_provinces_urban_ag_surpasses_vegetable_crop.html (accessed December 2, 2008).

2008c   "Desarrolla Cuba agricultura sobre una base ecológica, sostenible y familiar." September 7. http://www.cubaheadlines.com/es/2008/09/07/13308/desarrolla_cuba_agricultura_sobre_una_base_ecologica_sostenible_y_familiar.html (accessed December 2, 2008).

Cuba News Headlines

2008   "The University of Ciego de Avila." June 16. http://www.cubaheadlines.com/2008/06/16/11903/the_university_ciego_de_avila_five_enter_more_good_cuba.html (accessed March 14, 2009).

Cuesta Álvarez, Leonardo

2006   "Cuentas pendientes." *El Habanero Digital,* May 24. http://www.elhabanero.cubaweb.cu/2006/mayo/nro1600_may06/econ_06may570.html (accessed January 31, 2009).

De Bon, Herbert, Laurent Parrot, and Paule Moustier

2010   "Sustainable urban agriculture in developing countries." *Agronomy for Sustainable Development* 30(1): 21–32.

De Jesús, Lázaro

2008   "Sudando la gota gorda." *Granma,* December 19. http://www.granma.cubaweb.cu/2008/12/19/nacional/artic01.html (accessed December 19, 2009).

Delegación Provincial del CITMA, Comisión del Fórum

2005   "Llamamiento al Fórum XVI de Ciencia y Técnica." http://www.forumcyt.cu/forum/llamamiento (accessed February 26, 2009).

Delgado, Richard

2008   "Visita Ulises Rosales del Toro, la sede nacional de la ACTAF." *Agricultura Orgánica* 14(2): 21–22.

Desmarais, Anette Aurelie
2003   "Vía Campesina y la soberanía alimentaria." *La Jornada,* October 18.
Díaz Duque, José A., Nancy Machín Rodríguez, and Margarita Villar García
2004   "Los científicos en la batalla de ideas." *CIGET Pinar del Río* 6(1). http://www.
        citma.pinar.cu/ciget/No.2004-1/batalla%20de%20ideas.htm (accessed April 25,
        2009).
Díaz Herryman, Ana Laura
2009   "Universidades cubanas." Ministerio de Educación Superior. http://www.mes.
        edu.cu/index.php?option=com_content&task=view&id=125&Itemid=76 (ac-
        cessed March 28, 2009).
Díaz Otero, Soledad, and Emilio García Capote
2006   "Organización de la investigación y su infraestructura en el sector agrario." In
        Las investigaciones agropecuarias en Cuba: Cien años después, coord. María Te-
        resa Cornide Hernández, 1–21. Havana: Editorial Científico-Técnica.
Digital Granma Internacional
2008   "Biotecnología cubana: Aportes a la producción de alimentos." November 25.
        http://cuba-l.unm.edu/?nid=64019 (accessed February 25, 2009).
Ecoportal.net
2007   "La investigación científico-campesina en la agricultura cubana." July 11. http://
        www.ecoportal.net/content/view/full/70923 (accessed June 2, 2009).
EFE
2008   "Incentivan el pago para mejorar la productividad." *El Nuevo Herald,* June 23.
        http://www.elnuevoherald.com/noticias/america_latina/cuba/story/231599.
        html (accessed July 8, 2009).
Eleconomista.es
2007   "Raúl Castro impulsa perfeccionamiento empresarial para buscar eficiencia." Jan-
        uary 23. http://www.eleconomista.es/empresas-finanzas/noticias/135994/01/07/
        Raul-Castro-impulsa-perfeccionamiento-empresarial-para-buscar-eficiencia.
        html (accessed July 8, 2009).
2008   "Raúl Castro impulsa el 'pago por resultados' para aumentar la productividad
        laboral." June 22. http://www.eleconomista.es/economia/noticias/614102/06/08/
        Raul-Castro-impulsa-el-pago-por-resultados-para-aumentar-la-productividad
        -laboral.html (accessed July 8, 2009).
Elderhorst, Miriam
1994   "Will Cuba's biotechnology capacity survive the socioeconomic crisis?" *Bio-
        technology and Development Monitor* 20 (September: 11–13): 22. http://www.
        biotech-monitor.nl/2007.htm (accessed May 29, 2009).
El Portal Villa Clara
2009   "Empresa de Cultivos Varios." http://www.villaclara.cu/citma/sistema-ciencia/
        empresas-priorizadas/ecv (accessed March 19, 2009).
En Defensa de la Humanidad
2008   "Recomiendan en Cuba incrementar producción de alimentos." May 20. http://
        www.defensahumanidad.cult.cu/artic.php?item=6545 (accessed August 21,
        2008).

Espinosa Chepe, Oscar

2006   "La crisis de la producción agropecuaria cubana: Causas y posibles soluciones." *Cuba in Transition* 16: 14–23. http://lanic.utexas.edu/project/asce/pdfs/volume 16/pdfs/espinosachepe.pdf (accessed October 19, 2010).

Espinosa Martínez, Eugenio

1997   "The Cuban economy in the 1990s: From crisis to recovery." In *CartaCuba: Essays on the potential and contradictions of Cuban development*, ed. Richard A. Dello Buono and José Bell Lara, 9–20. Havana: FLACSO, Programa Cuba.

Espinosa Martínez, Odalys

2003   "Análisis y demandas tecnológicas: Base para la gestión del conocimiento y de la innovación en las empresas en perfeccionamiento empresarial." Paper presented at Infogest, October 23–24, Santiago de Cuba. http://bibliotecologia.udea.edu. co/elfaro/areas/even.html (accessed July 8, 2009).

Estrada Zamora, Raúl

2008   "Las Tunas, Cuba in search of an ecologically sustainable agrarian development." *Tiempo 21*, Trans. Ernesto Gutiérrez Pina. November 17. http://www. cubaheadlines.com/2008/11/20/14500/las_tunas_cuba_search_ecologically_ sustainable_agrarian_development.html (accessed December 2, 2008).

2008   "Las Tunas, Cuba propitiates more consumption of vegetables and condiments." *Tiempo 21*, August 29. http://www.tiempo21.cu/English/las_tunas/august08/ las_tunas_propitiates_more_consumption_vegetables_condiments_080829. htm (accessed December 2, 2008).

Facultad de Ciencias Agropecuarias de UCLV

2009   "Ciencia y técnica." http://www.agronet.uclv.cu (accessed February 27, 2009).

Farrell Villa, Juan

2007   "Inaugurarán nuevos módulos de casas de cultivo protegido y semiprotegido." *La Demajagua*, January 9. http://www.lademajagua.co.cu/infgran5625.htm (accessed August 21, 2008).

Febles Hernández, Miguel

2009   "No va lejos las de alante . . ." *Granma*, October 27. http://www.granma.cubaweb. cu/2009/10/27/nacional/artic01.html (accessed October 27, 2009).

Fernández, Elena María

2008   "En producción primera casa de cultivo semiprotegido en Güines." *Radio Güines Digital*, May 2. http://www.mayaweb.cu/noticia.asp?idLan=1&idNot=1685 (accessed July 23, 2008).

Figueroa Albelo, V. M., Ramón Sánchez Noda, Jaime García Ruiz, Antonio M. Ruiz Cruz, Julio R. Cárdenas Pérez, Maritza Victoria Martínez Lima, Marlene Penichet Cortiza, Orlando Saucedo Castillo, Grizel Donéztevez Sánchez, Ricardo Hernández Pérez, et al.

2006   *La economía política de la construcción del socialismo*. Electronic edition. http:// www.eumed.net/libros/2006b/vmfa/ (accessed July 8, 2009).

Fleitas Díaz, Mario, Jesús Mena Campos, Reinel Noa Ortega, Maykel Guzmán Castellanos, Eliberto Marrero, R. Dávila Vázquez, M. Fernández, et al.

n.d.   "Alternativa de sustitución del plaguicida químico BASAMID por el bioproducto 'HEBERNEM' en el control de *Meloidogyne incognita* chitwood." Working paper. Camagüey: Universidad de Camagüey, Centro de Ingeniería Genética y Biotecnología. http://docs.google.com/gview?a=v&q=cache:Fj6_hpNwJrwJ:cytdes. reduc.edu.cu/index.php%3Foption%3Dcom_docman%26task%3Ddoc_downl oad%26gid%3D224+mario+fleitas+diaz+BASAMID+HEBERNEM&hl=en&gl =us (accessed June 26, 2009).

Fleury, André, and Awa Ba

2005   "Multifunctionality and sustainability of urban agriculture." *Urban Agriculture Magazine* 15: 4–5.

Food and Agriculture Organization of the United Nations (FAO)

1995   *World agriculture: Towards 2010, an FAO study.* New York: FAO and John Wiley & Sons. Available online: FAO Corporate Document Repository. http://www. fao.org/docrep/v4200e/V4200E00.htm#Contents (accessed July 22, 2009).

2003   "Trade reforms and food security: Conceptualizing the linkages." Rome: FAO. Available online: FAO Corporate Document Repository. http://www.fao.org/ newsroom/en/news/2007/1000550/index.html (accessed March 15, 2010).

2006   "Food Security." Policy Brief 2, June.

2007   "Meeting the food security challenge through organic agriculture." FAO Newsroom. May 3. http://www.fao.org/newsroom/en/news/2007/1000550/index. html (accessed March 15, 2010)

2009   "Food security indicators: Cuba." *European Statistical System.* http://www.fao. org/fileadmin/templates/food security statistics/ country profiles/eng/Cuba E.pdf (accessed February 16, 2010).

FoodFirst News & Views

2007   "Food sovereignty and agroecology: Growing movements for constructive resistance." 29(107): n.p.

Food Research and Action Center (FRAC)

2008   "Low participation in food stamps means one in three struggling urban households miss out on benefits and hundreds of millions of federal dollars foregone." Press Release. http://www.frac.org/Press_Release/urbanfoodstamps08release. htm (accessed March 16, 2010).

Fórum de Ciencia y Técnica (FCT)

2008a   "Eventos del Fórum de Ciencia y Técnica." http://www.forumcyt.cu/forum/ eventos (accessed February 26, 2009).

2008b   "Cronología de los fórums nacionales." http://www.forumcyt.cu/fidel-y-el -forum/segunda-parte (accessed February 26, 2009).

Fresneda Buides, José A.

2006   "La producción de semillas en la agricultura urbana." *Agricultura Orgánica* 12(2): 36–37.

Fundora Mayor, Zoilá, ed.

2006   "27 de diciembre—Día de la Agricultura." Special issue, *Agricultura Orgánica* 12(2): 1–48.

Funes Aguilar, Fernando

2007   *Agroecología, agricultura orgánica y sostenibilidad.* Havana: Biblioteca ACTAF.

Funes Aguilar, Fernando, Luis García, Martin Bourque, Nilda Pérez, and Peter Rosset, eds.

2002   *Sustainable agriculture and resistance: Transforming food production in Cuba.* Oakland, Calif.: Food First Books.

Fuster Chepe, Eugenio

2006   "Diseño de la agricultura urbana cubana." *Agricultura Orgánica* 12(2): 6.

García, Adriano

1996   "La reestructuración industrial." *Cuba: Investigación Económica* April–June: 43–78.

García, Luis

1999   "Educación y capacitación agroecológica." *Agricultura Orgánica* 5(3): n.p.

García, Luis, Nilda Pérez, and E. Freyre

1999   "Centro de Estudios de Agricultura Sostenible: Su contribución a la difusión de la agricultura orgánica en Cuba." *Agricultura Orgánica* 5(3): 14–17.

García López, Libertad, and José Luís García Cueva

2006   "Investigación, formación de profesionales y estudios de postgrado en la educación superior agropecuaria en Cuba." In *Las investigaciones agropecuarias en Cuba: Cien años después,* coord. María Teresa Cornide Hernández, 22–57. Havana: Editorial Científico-Técnica.

Gasperini, Lavinia

2000   *The Cuban education system: Lessons and dilemmas.* Education Reform and Management Publication Series. Washington, D.C.: World Bank.

Gayoso, Antonio

2008   "An agricultural transition policy for Cuba: A message for Raúl." *Cuba in Transition* 18: 265–74. Available online: http://lanic.utexas.edu/project/asce/pdfs/volume18/pdfs/gayoso.pdf (accessed October 19, 2010).

Gobierno de Cuba

1992   "Decreto No. 175." *Gaceta Oficial de la República de Cuba* 90(13): 139.

Gómez Tovar, Laura, Manuel Ángel Gómez Cruz, and Rita Schwentesius Rindermann

2004   "Propuesta de política de apoyo para la agricultura orgánica de México." *Revista Electrónica Latinoamericana en Desarrollo Sustentable,* May 25. http://vinculando.org/organicos/apoyo_agricultura_organica.html (accessed March 23, 2010).

González, Rosalía, Yanel Ojeda, and J. L. Pozo

1999   "Impacto social de la agricultura urbana." *Agricultura Orgánica* 5(2): n.p.

González Díaz, Lianet, Miguel A. Hernández Estrada, Danneys Armario Aragón, Teresa Ramírez Pedraza, Jaime Simó González, Osvaldo Triana Martínez, Alfredo de la Nuez Figueroa et al.

n.d.   "Comportamiento productivo de 5 clones de plátano bajo el estudio de tres densidades de población." Instituto de Investigaciones de la Sanidad Vegetal. http://www.fao.org/docs/eims/upload/cuba/5388/PublicLianetFAOLianet.pdf (accessed July 3, 2009).

González Hernández, Gema

2005   "Prácticas agroecológicas: Desarrollo y limitaciones en las cooperativas (UBPC) Urbanas." Master's thesis, FLACSO, Universidad de la Habana.

González Martínez, Ortelio
1999   "Comenzó siembra de papa en Ciego." *Granma*, November 30. http://www.
        granma.cubaweb.cu/30nov99/nacional/articulo7.html (accessed November 30,
        1999).
González Novo, Mario
2006   "Una cooperativa del barrio para la ciudad." *Agricultura Orgánica* 12(2): 11–12.
Graña, Ángel
2005   "Historia de la Sociedad Espeleológica de Cuba: Memorias del VII Congreso
        Nacional Mexicano de Espeleología." *Montañismo y Exploración*. http://
        montanismo.org.mx/imprimir.php?id_sec=6&id_articulo=1155 (accessed
        March 30, 2009).
Granma
2007   "Villa Clara muestra su eficiencia en los cultivos." December 17. http://www.
        granma.cubaweb.cu/2007/12/17/nacional/artic18.html (accessed March 19,
        2009).
2009a   "Otorgan doble excelencia a organopónico de Cienfuegos." May 26. http://
        www.granma.cubaweb.cu/2009/05/26/nacional/artic07.html (accessed May 27,
        2009).
2009b   "Terminan 17 nuevos organopónicos de cultivos semiprotegidos." May 26. http://
        www.granma.cubaweb.cu/2009/05/26/nacional/artic09.html (accessed May 27,
        2009).
Grogg, Patricia
2008a   "Cuba: Local farmers producing food solutions." *Inter Press Service News Agen-
        cy*, June 5. http://ipsnews.net/news.asp?idnews=42673 (accessed December 2,
        2008).
2008b   "Un guardabosques cuida La Habana." *Inter Press Service News Agency*, Novem-
        ber. http://ipsnoticias.net/nota.asp?idnews=90551 (accessed October 16, 2009).
Grupo Nacional de Agricultura Urbana (GNAU)
2004   *Lineamientos para los subprogramas de la agricultura urbana para 2005–2007 y
        sistema evaluativo*. November. Havana: MINAG.
2007a   *Lineamientos para los subprogramas de la agricultura urbana para 2008–2010 y
        sistema evaluativo*. February. Havana: MINAG.
2007b   *Manual técnico de organopónicos y huertos intensivos*. March. Havana: MINAG.
Grupo Provincial de Agricultura Urbana (GPAU) de Matanzas
2007   *Información operativa de la agricultura urbana—Provincia Matanzas*. March.
        Matanzas: MINAG.
Grupo Provincial de Agricultura Urbana en la Ciudad de la Habana, Cuba
2006   "La agricultura urbana en los municipios de Ciudad de la Habana con enfoque
        sostenible y agroecológico." PowerPoint presentation. Havana. http://www.
        cipotato.org/urbanharvest/documents/pdf/La%20Habana%20-%20Orlando
        %20Acosta.pdf (accessed March 27, 2010).
Guerra Rey, Gabriela
2008   "Crisis alimentaria: Un reto de humanidad." *Bohemia*, July 23.

Guevara, Ernesto Ché

2003   *The Che Guevara reader.* New York: Ocean Press.

Hagelberg, G. B., and José Alvarez

2007   "Cuba's dysfunctional agriculture: The challenge facing the government." *Cuba in Transition* 17: 144–58. http://lanic.utexas.edu/project/asce/pdfs/volume17/pdfs/hagelbergalvarez.pdf (accessed October 19, 2010).

2009   "Cuban agriculture: The return of the campesinado." *Cuba in Transition* 19: 229–41. Available online: http://lanic.utexas.edu/project/asce/pdfs/volume19/pdfs/hagelbergalvarez.pdf (accessed October 19, 2010).

Halweil, Brian

2006   "Can organic farming feed us all?" *World Watch Magazine,* May–June. http://www.worldwatch.org/node/4060 (accessed March 15, 2010).

Heber Biotec

2007   "Áreas de cultivos protegidos tratados con HeberNem y plan." PowerPoint by Heber Biotec. http://74.125.47.132/search?q=cache:xy-9q9fEAsQJ:revistas.mes.edu.cu/eduniv/events/fitosanidad/HeberNem-Bionematicida-Ecologico.pdf/at_download/file+%22HeberNem+Bionematicida+ecologico:+producto+biologico+para+el+control+de+nematodos+parasitos+de+plantas%22&cd=1&hl=en&ct=clnk&gl=us (accessed July 21, 2009).

Hernández Díaz, María Isabel, Marisa Chailloux Laffita, and Anselma Ojeda Veloz

2006   "Cultivo protegido de las hortalizas: Medio ambiente y sociedad." *Temas de Ciencia y Tecnología* 10(30): 25–31.

Hernández Navarro, Luis

2008   "Vía Campesina: Reserva de futuro." *La Jornada,* October 28.

Hernández Pérez, Raúl

2006   "Consultorios tiendas agropecuarias (CTA), un eslabón imprescindible en los sistemas urbanos de producción." *Agricultura Orgánica* 12(2): 13.

Hoffmann, Heide

1999   "Urbane Landwirtschaft am Beispiel der Organopónicos in Havanna/Kuba." Paper presented at the Deutscher Tropentag conference, Berlin, October 14–15. http://ftp.gwdg.de/pub/tropentag/proceedings/1999/referate/STD_C11.pdf (accessed October 19, 2010).

Holt-Gimenez, Eric, and Raj Patel, with Annie Shattuck

2009   *Food rebellions! Crisis and the hunger for justice.* Oakland, Calif.: Food First Books.

Ibáñez López, Juan

2006   *La dialéctica productor directo—medios de producción: El periodo de la transición socialista en Cuba.* Electronic edition. http://www.eumed.net/libros/2006b/jil/ (accessed July 8, 2009).

Infante Curbelo, Leyanis

2008   "Naturaleza in vitro." *Ecohabanero,* March 19. http://www.elhabanero.cubasi.cu/eco/nro14/in-vitro.html (accessed May 29, 2009).

Instituto de Investigaciones de Sanidad Vegetal (INISAV)

2009   "Misión." http://www.inisav.cu/mision.htm (accessed March 14, 2009).
Instituto Nacional de Investigaciones de Viandas Tropicales (INIVIT)
2009   "Instituto Nacional de Investigaciones de Viandas Tropicales." Portal de Villa Clara. http://www.villaclara.cu/citma/polo/centros/inivit (accessed February 27, 2009).
Izquierdo, Juan, and Adolfo A. Rodríguez Nodals, eds.
2004   "Manual sobre agricultura orgánica sostenible." Oficina Regional de la FAO para América Latina y el Caribe. http://www.rlc.fao.org/es/agricultura/aup/pdf/organica.pdf (accessed July 22, 2009).
Jiménez Díaz, Alain
2008   "La Quinta: Portento de la agricultura urbana." Centrovisión, May 21. http://alainjd.blogia.com/2007/120401-la-quinta-portento-de-la-agricultura-urbana.php (accessed July 27, 2008).
Jiménez Terry, F. A., D. Ramírez Aguilar, and D. Agramonte Peñalver
2004   "Use of biobras-6 in micropropagation of FHIA-21." InfoMusa 13(1): 4–6. Available online http://bananas.bioversityinternational.org/files/files/pdf/publications/info13.1_en.pdf (accessed March 14, 2009).
Khor, Martin
2008   "Behind the July failure of WTO talks on Doha." Economic and Political Weekly, August 16.
Koont, Sinan
1994   "Cuba: An island against all odds." Monthly Review 46(5): 1–18.
2004   "Food security in Cuba." Monthly Review 55(8): 11–20.
2007   "Urban agriculture in Cuba: Of, by, and for the barrio." Nature, Society, and Thought 20(3–4): 311–26.
2010   "Urban agriculture in Cuba: Advances and challenges." Paper presented at the 29th International Congress of the Latin American Studies Association, Toronto, Canada, October 9.
Lage Dávila, Carlos
2007   "El perfeccionamiento empresarial en Cuba es un proceso de constante mejoramiento." Cuba Socialista, August 30. http://www.cubasocialista.cu/texto/009808llage.html (accessed July 8, 2009).
Lagnaoui, A.
2000   "A sustainable pest management strategy for sweetpotato weevil in Cuba: A success story." Food and Fertilizer Technology Center. December 1. http://www.agnet.org/library/eb/493a/ (accessed July 27, 2008).
Lavielle Laugart, Suraya
2002   "Nuevos paradigmas en el perfeccionamiento de la gestión empresarial en Cuba." Santiago de Cuba: Universidad de Oriente. http://www.uo.edu.cu/ojs/index.php/stgo/article/viewFile/14502442/662 (accessed July 15, 2009).
Lehmann, Volker
2000   "Cuban agrobiotechnology: Diverse agenda in times of limited food production." Biotechnology and Development Monitor 42 (June): 18–21. http://www.biotech-monitor.nl/4207.htm (accessed May 29, 2009).

Levins, Richard

2005 "How Cuba went ecological." *Revista Laberinto* 18(2). http://laberinto.uma.es/
index.php?option=com_content&view=article&id=289:how-cuba-went-ecolog
ical&catid=52:lab18&Itemid=54 (accessed June 30, 2009).

Lotti, Alina M.

2008 "Crecerá matricula en institutos politécnicos agropecuarios." *Trabajadores,*
July 21. http://www.trabajadores.cu/news/crecera-matricula-en-institutos-
politecnicos-agropecuarios (accessed June 21, 2009).

MacEwan, Arthur

1981 *Revolution and economic development in Cuba.* New York: St. Martin's Press.

Maestri, Nicole, and Lisa Baertlein

2009 "Midnight in the food-stamp economy." Reuters News Agency. December 18.
http://www.reuters.com/assets/print?aid=USTRE5BH2C220091218 (accessed
March 16, 2010).

Manuel Castello, Luis

2008 "Las ciudades y sus entornos verdes y saludables." *Rebelión,* July 30. http://www.
rebelion.org/noticia.php?id=70897 (accessed July 30, 2008).

Martín, Lucy

1999a "Aspectos económicos y sociales en dos faros agroecológicos." *Agricultura
Orgánica* 5(2): 38–41.

1999b "Punto por punto: Aspectos económicos y sociales en dos faros agroecológicos."
Agricultura Orgánica 5(2): n.p.

Martínez Martínez, Osvaldo, Daniel Codorniú Pujals, and Miguel Márquez

2004 "Investigación sobre ciencia, tecnología y desarrollo humano en Cuba 2003."
United Nations Development Programme and El Centro de Investigaciones de
la Economía Mundial.

Martínez Rodríguez, Francisco

2006 "Abonos orgánicos y su contribución a la sostenibilidad de los sistemas agrícolas
en Cuba." *Agricultura Orgánica* 12(2): 40–42.

Martínez Zubiaur, Yamila

2007 "Enfermedades emergentes en sistemas de cultivos protegidos." PowerPoint presentation at the Primero Seminario Técnico de Cultivos Protegidos, Havana,
September 25–28.

Martín García, Marili, and Enrique Rodríguez Corominas.

n.d. "El costo de producción en procesos de micropropagación para biofábricas de
múltiples cultivos." Universidad de Las Villas, Villa Clara.

Martín González, Marielena, and Dora Pérez Sáez

2008 "Por un cambio en la comercialización agropecuaria." *Juventud Rebelde,* January
27.

Massip, José A., Ernesto Hernández García, and Boris Nerey Obregón

2001 "La empresa estatal cubana y el Proceso de Perfeccionamiento Empresarial." *La
Revista Cubana de Ciencias Sociales* 32. http://www.nodo50.org/cubasigloXXI/
economia/massip_hdez_nerey1_230101.htm (accessed July 8, 2009).

Mayoral, María Julia

2007   "Recorre Raúl cooperativa agropecuaria en Pinar del Río con alentadores resultados: Testimonio de gente decidida a trabajar la tierra. Producen todo lo que se proponen." *Granma Digital,* March 15. http://granmai.cubasi.cu/espanol/2007/marzo/juev15/recorre.html (accessed July 23, 2008).

Medina Basso, Nicolás, and Ondina León Díaz

2004   *Biotechnology and sustainable agriculture: Biofertilizers and biopesticides.* PowerPoint presentation at the second workshop of Pugwash Meeting No. 294: The Impact of Agricultural Biotechnology on Environmental and Food Security, Havana, April 1–4. http://www.scribd.com/doc/4679348/Biofertilizer (accessed March 27, 2010)

Mejías Osorio, José

2007   "De Campesino a Campesino agroecología eficiente." *Solvisión Digital,* June 26. http://www.solvision.co.cu/PT/CYTcampesino/campesino.htm (accessed July 23, 2008).

Mena Campos, Jesús

2005   "Manual de aplicación del bionematicida HeberNem." Camagüey: CIGB.

Mena Campos, Jesús, Eulogio Pimentel Vázquez, Armando Hernández García, Licette León Barreras, Yamilka Ramírez Núñez, Idania Wong Padilla, Marieta Marín Bruzos, et al.

2007   "HeberNem: Sustituto de tratamientos químicos en Cultivos Protegidos." Working paper, Centro de Ingeniería Genética y Biotecnología (CIGB), Camagüey. http://www.forumcyt.cu/UserFiles/forum/Textos/0908001.pdf (accessed June 26, 2009).

Michelena, José Antonio

2009   "El patio de mi casa es comunitario: Una filosofía hacia la naturaleza." *Cultura y Sociedad* 1 (January). http://cubaalamano.net/sitio/print/article.php?id=10821 (accessed June 30, 2009).

Miguel, José

2007   "Julio: Un hombre productivo." José Miguel Online. November 16. http://josmiguelonline.blogspot.com (accessed June 9, 2009).

Milanés León, Enrique

2005   "New biological product for agriculture." *Granma,* December 15. http://granmai.cubasi.cu/ingles/2005/diciembre/juev15/51camaguey.html (accessed July 23, 2008).

Milián, Marilis, I. Sánchez, Magali García, S. Rodríguez, D. Guerra, J. M. Porieles, M. Hernández, Amparo Corrales, M. S. Lago, J. García, and María Oliva

n.d.   "Bromatología, caracteres culinarios y de resistencia a los ácaros en nuevos clones de malanga isleña (*Colocasia esculenta* Schott.) en Cuba." Instituto de Investigaciones en Viandas Tropicales (INIVIT). http://www.villaclara.cu/UserFiles/File/Portal%20prov./RInvest.doc (accessed March 17, 2009).

Milián Salaberri, Elena

2009   "Desarrollo arrocero ocupa a científicos vueltabajeros." *TelePinar,* March 30.

http://www.telepinar.icrt.cu/index.php?option=com_content&task=view&id
=5410&Itemid=38 (accessed April 30, 2009).

Ministerio de Ciencia, Tecnología y Medio Ambiente (CITMA)

2009　"Universidad agraria de La Habana (UNAH) 'Fructuoso Rodríguez.'" http://
www.citmahabana.cu/centr_c_unah.htm (accessed March 17, 2009).

Ministerio de Educación de Cuba

2009　"Técnica y profesional: Instituto Politécnico Agropecuario." Portal Educativo
Cubano.　http://www.cubaeduca.rimed.cu/index.php?option=com_content
&task=blogcategory&id=18&Itemid=37 (accessed June 21, 2009).

Ministerio de Educación Superior (MES)

2008　"Boletín estadístico." http://www.mes.edu.cu/Documentos/riaces/Boletín_
bolsillo08.pdf (accessed March 27, 2010)

2009　"Historia universitaria." http://www.mes.edu.cu/index.php?option=com_con
tent&task=view&id=12&Itemid=27 (accessed March 28, 2009).

Ministerio de Educación Superior, Junta de Acreditación Nacional (MES JAN)

2009　"Definition of Key Terms." http://www.mes.edu.cu/index.php?option=com_con
tent&task=view&id=66&Itemid=5 (accessed March 24, 2010).

Ministerio de la Agricultura (MINAG), Grupo Técnico de Bio-fábricas y Plátano

2004　"Tecnología de futuro: Una nueva concepción en la producción de plátano fruta
y vianda en Cuba). Havana: MINAG.

Ministerio de Relaciones Exteriores

2008　"Cambio climático y biodiversidad." Temas Multilaterales. http://www.cuba
minrex.cu/Multilaterales/Articulos/Archivo/2008/0606-Cambio2.html (accessed October 17, 2009).

Monteagudo, Katia

2009　"Sacarle el jugo a la tierra." *Bohemia,* January 23. http://www.bohemia.cubasi.
cu/2009/01/23/encuba/agricultural.html (accessed February 24, 2009).

Morales, Esteban

2008　"Algunos antecedentes históricos: El conflicto Cuba-EEUU desde el umbral del
siglo XXI." *La Jiribilla,* December 27, 2008.

Mougeot, Luc J. A.

1994　*The Asian leadership: City-farming today as it will be tomorrow.* City Farmer
Urban Agricultural Notes. http://www.cityfarmer.org/lucAsialead30.html (accessed September 12, 2009).

2000　"Urban agriculture: Definition, presence, potential and risks." In *Growing cities,
growing food: Urban agriculture on the policy agenda,* ed. N. Bakker, M. Dubbeling, S. Gundel, U. Sabel-Koschella, and H. de Zeeuw, 1–42. Feldafing, Germany:
German Foundation for International Development (DSE).

2006　*Growing better cities: Urban agriculture for sustainable development.* Ottawa:
IDRC.

Mougeot, Luc J. A., ed.

2005　*Agropolis: The social, political, and environmental dimensions of urban agriculture.* Ottawa: IDRC/Earthscan.

Muiño, Berta L., Eleazar Botta, Adriana Ballester, Eduardo Pérez, Emilio Fernández, Davis Moreno, and Ricardo Cuadras

n.d. "Manejo de plagas y enfermedades: Alternativas al bromuro de metilo." PowerPoint presentation. http://74.125.95.132/search?q=cache:N2pDUPalz34J:revistas.mes.edu.cu/eduniv/events/fitosanidad/Alternativas-Bromuro-Metilo.pdf/at_download/file+manejo+de+plagas+y+enfermedades+alternativas+al+bromuro+de+metilo&cd=2&hl=en&ct=clnk&gl=us (accessed July 1, 2009).

Muñoz, Eulogio V., Ángel González, Emigdio Rodríguez, and José Zambrano

2004 "Informe final de monitoreo: Escalonamiento, capacitación, y difusión de experiencias exitosas en la agricultura con principios agroecológicos en Cuba." Habana: Asociación Cubana de Técnicos Agrícolas y Forestales (ACTAF). http://idl-bnc.idrc.ca/dspace/handle/123456789/26722 (accessed June 2, 2009).

Muñoz Pérez, Naily, and María T. Mojáiber

2009 "Más eficacia en producción de bionematicida cubana." *Digital Granma Internacional,* February 17. http://www.granma.cu/espanol/2009/febrero/mar17/bionematicida.html (accessed February 25, 2009).

Murphy, Catherine

1998 *Cultivating Havana: Urban agriculture and food security in the years of crisis.* Development Report #12. Oakland, Calif.: Institute for Food and Development Policy.

Navarro, Medrado

2007 "Manejo orgánico para cultivos protegidos y semiprotegidos." PowerPoint presentation at the Primer Seminario Técnico de Sanidad Vegetal en Casas de Cultivo Protegido, Havana, September 25–28.

Nelson, Erin, Stephanie Scott, Julie Cukier, and Ángel Leyva Galán

2008 "Institutionalizing agroecology: Success and challenges in Cuba." *Agriculture and Human Values* 26(3): 233–44.

Nova González, Armando

1999 "Cuba: Transformaciones de su sistema agroproductivo." *Agricultura Orgánica* 5(2): n.p.

2006 *La agricultura en Cuba: Evolución y trayectoria (1959–2005).* Havana: Editorial de Ciencias Sociales.

2008a "El actual mercado interno de los alimentos." *Cuba Siglo XXI,* July. http://www.nodo50.org/cubasigloXXI/economia/nova2_300608.pdf (accessed July 8, 2009).

2008b "La agricultura en Cuba: Actualidad y transformaciones necesarias." *Cuba Siglo XXI,* August. http://www.nodo50.org/cubasigloXXI/economia/novag_310808.pdf (accessed July 8, 2009).

2008c "La economía cubana y las fuentes alternativas de energía renovable." *Cuba Siglo XXI,* August. http://www.nodo50.org/cubasigloXXI/economia/novag5_310808.pdf (accessed July 8, 2009).

2008d "Importancia económica del sector agropecuario en Cuba." *Cuba Siglo XXI,* November. http://www.nodo50.org/cubasigloXXI/economia/nova_311008.pdf (accessed July 8, 2009).

2008e   "El modelo agrícola cubano en la etapa 1993–2008." *Cuba Siglo XXI*, August. http://www.nodo50.org/cubasigloXXI/economia/novag4_310808.pdf (accessed July 8, 2009).

2008f   "Modelo agrícola cubano surgimiento y evolución 1510 hasta 1959." *Cuba Siglo XXI*, July. http://www.nodo50.org/cubasigloXXI/economia/nova_300608.pdf (accessed July 8, 2009).

2008g   "El modelo de desarrollo agrícola cubano en el periodo 1959–1990." *Cuba Siglo XXI*, August. http://www.nodo50.org/cubasigloXXI/economia/novag3_310808. pdf (accessed July 8, 2009).

2008h   "La necesidad de un modelo agrícola eficiente." *Cuba Siglo XXI*, August. http:// www.nodo50.org/cubasigloXXI/economia/novag2_310808.pdf (accessed July 8, 2009).

Novo Sordo, René, and Juan Germán Hernández Barrueta
2009   *Historia de la microbiología del suelo en Cuba.* Havana: Editorial Universitaria.

Núñez Jiménez, A.
1964   "Consideraciones en torno a la revolución científico-técnica en Cuba." *Cuba Socialista*, 1st ser., 38 (October): 44–56.

NutriNet Cuba
2009   "PMA en acción." http://cuba.nutrinet.org/servicios/zona-infantil/lpma-en -accionr (accessed June 20, 2009).

Oficina Nacional de Estadísticas de Cuba
2009   *Anuario estadístico de Cuba 2008—eEdición 2009.* http://www.one.cu/aec2008 (accessed September 12, 2009).

Ojeda, Janet, and Nelson Companioni Concepción
1999   "Capacitación: Herramienta básica para el desarrollo de la agricultura urbana." *Agricultura Orgánica* 5(3): n.p.

Oramas León, Orlando
2008   "Cosechar en una hectárea lo que produce una caballería." *Granma*, October 24. http://www.granma.cubaweb.cu/2008/10/24/nacional/artic01.html (accessed January 31, 2009).

O'Reilly Morris, Elvira
2006   "Adopción de medios biológicos para el control de plagas en el sector cooperativo campesino." Master's thesis, FLACSO, Universidad de la Habana.

Pablo Sáez, Pedro
2007   "Ciencia y técnica en función de la alimentación del pueblo." *Televisión Camagüey*, August 23. http://www.tvcamaguey.co.cu/x/2007/noticias/ampli-ar/5817/index.htm (accessed February 26, 2009).

Pagés, Raisa
2006a   "Agricultura urbana en 2006." *Granma*, December 31.
2006b   "Una ciudad agroecológica." *Agricultura Orgánica* 12(2): 16–18.
2007a   "La agricultura capitalina se deshace del marabú." *Granma*, March 3.
2007b   "Mas hortalizas, pero el aporte de algunos territorios no es suficiente." *Granma*, July 26. http://www.granma.cubaweb.cu/2007/07/26/nacional/artic02.html (accessed July 23, 2008).

2007c   "Pese a clima adverso, producción hortícola se mantuvo." *Granma,* May 20. http://www.granma.cubaweb.cu/2007/05/20/nacional/artic01.html (accessed July 23, 2008).

2008a   "Prácticas agroecológicas ante altos precios de insumos." *Granma,* May 15. http://www.granma.cubaweb.cu/2008/05/15/nacional/artic04.html (accessed May 15, 2008).

2008b   "Reacción rápida ante las dificultades del clima." *Granma,* January 16. http://www.granma.cubaweb.cu/2008/01/16/nacional/artic01.html (accessed July 23, 2008).

Palomares Calderón, Eduardo

2008   "Profundizan integración entre polos científicos: Onstalan equipos de electrocardiografía en el Hospital Juan Bruno Zayas, de Santiago de Cuba." *Trabajadores,* January 15. http://www.trabajadores.cu/news/cuba-enero-2008/pro fundizan-integracion-entre-polos-cientificos/ (accessed April 16, 2009).

Pan American Health Organization

1999   *Cuba: Profile of the Health Services System.* Washington, D.C.: 1–23.

Peláez, Orfilio

2009   "Espacios verdes para la ciudad." *Granma,* October 16.

Peralta García, Esther L., Luis Pérez Vicente, Orietta Fernández-Larrea, Gonzalo Dierksmeier Corcuera, and María E. Rodríguez Fuentes

2006   "Panorámica del desarrollo de la sanidad vegetal en Cuba: Sus principales hitos y resultados." In *Las investigaciones agropecuarias en Cuba: Cien años después,* coord. María Teresa Cornide Hernández, 222–44. Havana: Editorial Científico-Técnica.

Pérez, Nilda

1997   "Bioplaguicidas y agricultura orgánica." *Agricultura Orgánica* 3(2): n.p.

2008   "UBPC 'El Mango': Doble Corona." *Agricultura Orgánica* 14(2): n.p.

Pérez Betancourt, Roberto

2008   "Revitalization of Cuban urban and traditional agriculture." *Periódico 26,* July. http://www.periodico26.cu/english/news_cuba/july2008/cuba-agricul ture071908.html (accessed March 11, 2009).

Pérez Cabrera, Freddy

2008   "La agricultura en Villa Clara sustituye importaciones." *Granma,* August 16. http://www.granma.cubaweb.cu/2008/08/16/nacional/artic06.html (accessed August 16, 2008).

2009   "La hora de Acopio." *Granma,* January 30. http://www.granma.cubaweb. cu/2009/01/30/nacional/artic01.html (accessed January 30, 2009).

Pérez Consuegra, Nilda

2002   "Agricultura orgánica: Una visión desde Cuba." *Agricultura Orgánica* 8(2): 6–11.

Pérez Cruz, Felipe de J.

2009   "En el 50 aniversario de la Revolución Cubana: La ciencia y la tecnología como decisión trascendental." *Cuba Socialista,* January 16. http://www.cubasocialista. cu/texto/985948frf.html (accessed January 17, 2009).

Pérez Navarro, Lourdes

2008 "Cuba: nuevo sistema de pago por resultados." *Cubainformación*, June 12. http://www.cubainformacion.tv/index.php?option=com_content&task=view&id=546 2&Itemid=65 (accessed July 8, 2009).

Pérez Peñaranda, María Cristina

2001 "Formación de recursos humanos en biotecnología en Cuba." REDBIO/FAO. http://www.redbio.org/portal/encuentros/enc_2001/mesaredonda/M-06/ Mar%EDa%20Cristina%20P%E9rez/M-06.pdf (accessed July 3, 2009).

Pérez Sáez, Dora

2006 "Exporta Cuba diseño de instalaciones de cultivo protegido." *Juventud Rebelde*. Exportapymes, October 17. http://www.exportapymes.com/comercio-exterior -america-esp/exporta-cuba-diseno-de-instalaciones-de-cultivo-protegido (accessed July 22, 2009).

2008 "El 15 de enero se conmemora el Día de la Ciencia en Cuba." *Juventud Rebelde*, April 14.

PGU-ALC Habitat Programa de las Naciones Unidas para el Desarrollo Programa de Gestion Urbana

2000 Declaración de Quito. Final declaration of the Agricultura Urbana en las Ciudades del Siglo XXI workshop-seminar, Quito, Ecuador, April 10–20.

Pinareño, Agro

2009 "Las aguas vuelven a su cauce." *Bohemia*, January 23. http://www.bohemia.cu-basi.cu/2009/01/23/encuba/agricultura.html (accessed February 24, 2009).

Pollan, Michael

2007 *The omnivore's dilemma: A natural history of four meals*. New York: Penguin Press.

Pradas, Toni

2004 "INIFAT: El huerto cumple cien años." *Bohemia Digital*, May. http://www.bohemia.cubaweb.cu/2004/05/01SEMANA/sumarios/cienciatecnologia/articulo1.html (accessed June 29, 2009).

Prensa Latina

2008 "La Habana preparada para llegada de huracán Ike."September 8. http://www.prensa-latina.com.ar/article.asp?ID=%7BBEDFD053-0A9D-4EB3-81EC-3F1A2 9DFAD4A%7D) (accessed September 8, 2008).

Puente Nápoles, José

2006 "La agricultura urbana asume el abastecimiento de hortalizas a círculos infan-tiles, escuelas y hospitales." *Agricultura Orgánica* 12(2): 22.

Quevedo Rodríguez, Vito

2006 "Evolution and current state of science, technology, and innovation: Republic of Cuba." PowerPoint presentation at the International Conference of S&T Policy Research and Statistical Indicators, November 8–10, Colombo, Sri Lanka.

Quintana Martínez, Onexy

n.d. "La agricultura como vía de desarrollo en Cuba." *Monografías*. http://www.monografias.com/trabajos36/agricultura-cuba/agricultura-cuba.shtml (accessed July 8, 2009).

Rebelión.org
2008a   Carta de Fidel Castro al VII Congreso de la UNEAC 3 April. http://www.
         rebelion.org/noticias/2008/4/65506.pdf (accessed April 3, 2008).
2008b   "Soberanía alimentaria y Revolución: Aprendamos de Tachai." December 30.
         http://www.rebelion.org/noticia.php?id=78153 (accessed December 30, 2008).
Rendón Matienzo, Fidel
2009   "II encuentro de círculos de interés agropecuarios y forestales." Tribuna de La
         Habana, May 13. http://www.tribuna.co.cu/etiquetas/2009/mayo/13/encuentro
         -circulos.html (accessed June 30, 2009).
Revista.mes.edu.cu
2007   "Relatoría del Primer Seminario Técnico de Sanidad Vegetal en casas de cultivo
         protegido." http://www.revista.mes.edu.cu/eduniv/evento/fitosanidad/Relatorio
         -Seminario-Fitosanidad.pdf (accessed July 27 2009).
Rey Veitía, Lourdes
2006   "Amplio desarrollo de vitroplantas en Cuba." Trabajadores, May 11. http://edicion
         esanteriores.trabajdores.cu/2006/mayo/11/cuba/lr-vitroplantas.htm (accessed
         March 14, 2009).
2008   "Renace Valle del Yabú." Trabajadores, August 13. http://www.trabajadores.cu/
         news/renace-valle-del-yabu-1 (accessed August 13, 2008).
Riquenes Cutiño, Odalis
2008   "Santiago de Cuba está inmersa en una transformación en todos los órdenes."
         Juventud Rebelde, July 15. http://www.juventudrebelde.cu/cuba/2008-07-15/
         santiago-de-cuba-esta-inmersa-en-una-transformacion-en-todos-los-ordenes/
         (accessed July 23, 2008).
Rodríguez, Luis C., and Hermann M. Niemeyer
2005   "Integrated pest management, semiochemicals and microbial pest-control
         agents in Latin American agriculture." Crop Protection 24(7): 616–22.
Rodríguez Castellón, Santiago
2003   "La agricultura urbana y la producción de alimentos: La experiencia de Cuba."
         Cuba Siglo XXI 30 (June): 77–101. http://www.nodo50.org/cubasigloXXI/econo
         mia2.htm (accessed July 23, 2009).
Rodríguez Cruz, Francisco, Ramón Brizuela Roque, Lourdes Rey Veitía, Orestes Ramos
         Lorenzo, and Moisés González Yero
2003   "Polos que atraen." Trabajadores, May 12. http://edicionesanteriores.traba
         jadores.cu/2003/mayo/12/ciencias/polos.htm (accessed April 16, 2009).
Rodríguez González, Manuel
2008   El abc del productor Cubano. Digital monograph. http://www.monogafis.com/
         trabajos59/productor-agricola-cubano/productor-agricola-cubano.shtml (ac-
         cessed August 31, 2009).
Rodríguez Nodals, Adolfo A.
2006   Síntesis histórica del movimiento nacional de agricultura urbana de Cuba. Agri-
         cultura Orgánica 12(2): 26–27.
2007   La organoponía semiprotegida: Una tecnología de futuro para el trópico. Power-
         Point presentation at the Primer Seminario Técnico de Sanidad Vegetal en Casa

de Cultivos Protegidos. http://revistas.mes.edu.cu/eduniv/events/fitosanidad/ Organoponia-Semiprotegida.pdf/view (accessed July 1, 2009).

Rodríguez Nodals, Adolfo A., and Nelso Companioni Concepción

2006   "Situación actual: Perspectivas y retos de la agricultura urbana en Cuba." *Agricultura Orgánica* 12(2): 4–5.

Rodríguez Nodals, Adolfo A., Nelso Companioni Concepción, and Rosalía González Bayón

2006   "La agricultura urbana y periurbana en Cuba: Un ejemplo de agricultura sostenible." PowerPoint presentation at the VI Encuentro de Agricultura Orgánica, Havana.

Rodríguez Nodals, Adolfo A., Nelso Companioni Concepción, and Maria Elena Herrería Martínez

2006   "Las granjas urbanas en la agricultura cubana." *Agricultura Orgánica* 12(2): 7–9.

Royce, Frederick S.

2004   "Agricultural production cooperatives in Cuba: Toward sustainability." *Cuba in Transition* 14: 254–73. Available online: http://lanic.utexas.edu/project/asce/pdfs/volume14/royce.pdf (accessed October 19, 2010).

Sáez Chávez, Armando, and Ramón Barreras Ferrán

2008   "Marcha sostenida de la agricultura urbana en Cienfuegos." *5 Septiembre,* March 3. http://www.5septiembre.cu/agricultura148.htm (accessed July 23, 2008).

Salcines López, Miguel Ángel, and Isis María Salcines Milla

2007   "Fomento y desarrollo de una cooperativa de producción orgánica: UBPC Organopónico Vivero Alamar." PowerPoint presentation in possession of Miguel Salcines, Havana.

Sánchez, L., and L. Chirino

1999   "De Campesino a Campesino: Apuntes para una propuesta." *Agricultura Orgánica* 5(3): 26–29.

Sánchez Balboa, Modesto

2006   "Las fincas estatales en la agricultura urbana." *Agricultura Orgánica* 12(2): 10.

Santa Cruz, Gladys, and Mercedes Mayarí

1999   "Agricultura y desarrollo rural sostenible en los institutos politécnicos agropecuarios." *Agricultura Orgánica* 5(3): 19–20.

Santos Riveras, América

2008   "El sistema de ciencia e innovación tecnológica: Herramientas de gestión." PowerPoint presentation at the Taller Proyección y Prospección en TIC, Havana, October 15.

Sanz Araujo, Lucía

2009   "En Cuba: Ciencia para el desarrollo." *Radio Rebelde,* January 12. http://www.radiorebelde.com.cu/noticias/ciencia/ciencia1-120109.html (accessed April 16, 2009).

Sanz Medina, Mireya

2001   *La agricultura urbana en Cuba: Un movimiento de popularización por la seguridad alimentaria.* October. Havana: World Data Research Center.

Saucedo, Milvia, and Osliani Figueiras
2009   "Las probadas ventajas del perfeccionamiento empresarial." Telecentro de la Provincia Sancti Spíritus, July 5. http://www.centrovision.cu/Enmarcha/probadas100609.htm (accessed July 8, 2009).

Schiøler, Ebbe
2000   "Annual Report, 2000." Centro Internacional de la Papa. http://www.cipotato.org/publications/annual_reports/2000/05.asp (accessed July 27, 2008).

Siegelbaum, Portia
2008   "Cuba's urban agrarians flourish." CBS News, June 4. http://www.cbsnews.com/stories/2008/06/04/world/main4154650.shtml (accessed March 27, 2010).

Sierra, Raquel
2009a   "ACTAF puede contribuir a transformación agropecuaria." Tribuna de La Habana, February 2. http://www.tribuna.co.cu/etiquetas/2009/febrero/2/ACTAF-agropecuaria.html (accessed February 5, 2009).

2009b   "Campesinos incrementan producción de leche." Tribuna de La Habana, February 4. http://www.tribuna.co.cu/etiquetas/2009/febrero/4/campesinos-incrementan.html (accessed February 5, 2009).

2009c   "Comienza III Congreso de Asociación de Técnicos Agrícolas y Forestales." Tribuna de La Habana, February 2. http://www.tribuna.co.cu/etiquetas/2009/febrero/2/comienza-congreso-agr%C3%ADcolas-forestales.html (accessed February 5, 2009).

2009d   "Concluye III Congreso de la Asociación Cubana de Técnicos Agrícolas y Forestales." Tribuna de La Habana, February 4. http://www.tribuna.co.cu/etiquetas/2009/febrero/4/concluye-congreso.html (accessed February 5, 2009).

2009e   "Desarrolla la ciudad programa bufalino." Tribuna de La Habana, January 28. http://www.tribuna.co.cu/etiquetas/2009/enero/28/desarrolla.html (accessed February 5, 2009).

2009f   "En las transformaciones agropecuarias, vital la capacitación." Tribuna de La Habana, February 3. http://www.tribuna.co.cu/etiquetas/2009/febrero/3/transformaciones.html (accessed February 5, 2009).

2009g   "Proteger y usar el bosque." Tribuna de La Habana, April 15. http://www.tribuna.co.cu/etiquetas/2009/abril/15/proteger-bosque.html (accessed February 5, 2009).

Silva, Lázaro
2006   "35 años de la horticultura cubana." Cadena Habana, October 12. http://www.cadenahabana.cu/chabana2006/noticias/provinciales/provinciales01121006.htm (accessed July 13, 2009).

Simón, Félix
2006   "Campesinos científicos?" Campesino Cubano. Published by ANAP. November 29. http://www.campesinocubano.anap.cu/2006/noviembre/29novagroecologia.htm (accessed March 25, 2010).

Singh Castillo, Manuel de Jesús
2009   "Una iniciativa ambientalista encomiable." Solvisión Digital, May 25. http://

www.solvision.co.cu/Espanol/PT/iniciativa_ambientalista_0210509.html (accessed June 7, 2009).

Smit, Jac, and Martin Bailkey

1996 "Urban agriculture and the building of communities." In *Cities farming for the future: Urban agriculture for green and productive cities,* ed. René van Veenhuizen, 145–70. Ottawa: IDRC. Available online: http://www.idrc.ca/en/ev-1037 77-201-1-DO_TOPIC.html (accessed October 19, 2010).

Smith, Aaron

2010 "1 in 7 Americans rely on food stamps." CNNMoney.com, December 21. http://money.cnn.com/2010/12/21/news/economy/food_stamps/index.htm (accessed March 19, 2011).

Solberg, Ronald

2008 "US on reverse socialism path." *Asia Times Online,* September 25. http://www.atimes.com/atimes/Global_Economy/JI25Dj02.html (accessed September 24, 2008).

Spiaggi, Eduardo

2005 "Urban agriculture and local sustainable development in Rosario, Argentina: Integration of economic, social, technical and environmental variables." In *Agropolis: The social, political, and environmental dimensions of urban agriculture,* ed. Luc J. A. Mougeot, 187–99. Ottawa: IDRC/Earthscan.

Spieldoch, Alexandra

2008 "The food crisis and global institutions." Foreign Policy in Focus, August 6. http://www.commondreams.org/archive/2008/08/06/10862/print/ (accessed August 6, 2008)

StateMaster.com

2010 "Estimated number of illegal immigrants (most recent) by state." http://www.statemaster.com/graph/peo_est_num_of_ill_imm-people-estimated-number-illegal-immigrants (accessed March 16, 2010).

Suárez Pérez, Eugenio

2008 "La crisis alimentaria: Una realidad inobjetable." *Cuba Socialista,* August 19.

Suárez Ramos, Ronal

1999 "Avanza proceso de balance de la ANAP previo a su IX Congreso." *Granma,* November 27. http://www.granma.cubaweb.cu/27nov99/nacional/articulo8.html (accessed November 29, 1999).

2007a "Aumentaran cultivos semiprotegidos." *Granma,* February 12.

2007b "Frutos de la ciencia aplicada al tabaco." *Granma,* February 10. http://www.granma.cubasi.cu/secciones/ciencia/ciencia360.htm (accessed April 30, 2009).

Suárez Rivas, Ronald, and Ronal Suárez Ramos

2009 "La triple corona de El Mango." *Granma Digital,* June 6. http://www.granma.cubaweb.cu/2009/06/06/nacional/artic03.html (accessed June 6, 2009).

Taset Aguilar, Mariagny

2005 "La sabiduría se vuelve ciencia." *El Habanero,* January 15. http://www.elhabanero.cubaweb.cu/2005/enero/nro1176_05ene/cienc_05ene317.html (accessed on June 2, 2009).

TeleSur
2008   "Ministros de Petrocaribe crean fondo alimentario de 450 millones de dólares."
        July 30. http://www.aporrea.org/imprime/n117899.html (accessed July 31, 2008).

Terrero, Ariel
2009   "Neuronas para una economía intranquila." *Bohemia,* January 9. http://www.
        bohemia.cubasi.cu/2009/01/09/economia/economia.html (accessed February
        15, 2009).

Tharamangalam, Joseph
2008   "Can Cuba offer an alternative to corporate control over the world's food sys-
        tem?" Paper presented at the twentieth Conference of North American and Cu-
        ban Philosophers and Social Scientists, Havana, June. http://www.globaljustice
        center.org/articles/report_cubafood.html (accessed July 7, 2009).

Treminio, Reynaldo
2004   "Experiencias en agricultura urbana y peri-urbana en América Latina y el Ca-
        ribe: Necesidades de políticas e involucramiento institucional." Working docu-
        ment of RLCP/TCA, No. 001. Santiago de Chile: FAO, Regional Office for Latin
        America and the Caribbean.

Tristá Arbesú, Grisel
2000   "El perfeccionamiento empresarial en la empresa estatal cubana y socialista:
        Momentos actuales." *Cuba Socialista* 18: n.p.

Ubieta Gómez, Raimundo
2008   "IP policy at the Cuban biotechnology." PowerPoint presentation at the Life Sci-
        ences Symposium, Geneva, December 15. http://www.wipo.int/meetings/en/
        doc_details.jsp?doc_id=114637 (accessed July 2, 2009).

United Nations Environment Programme (UNEP)
2004   "Methyl bromide." *OzonAction* 48 (September): n.p.

Urban, Francis
1991   "World fertilizer use continues up—current trends." *World Agriculture,* June.
        Available on BNET http://findarticles.com/p/articles/mi_m3809/is_n63/ai
        ?11173964 (accessed July 22, 2009).

Urquhart, Sam
2008   "Food crisis, which crisis?" *Z Magazine,* July.

Valdés, Héctor, coord.
2008   *Los aprendizajes de los estudiantes de América Latina y el Caribe: Primer reporte
        de los resultados del Segundo Estudio Regional Comparativo y Explicativo.* Oficina
        Regional de Educación de la UNESCO para América Latina y el Caribe. San-
        tiago, Chile: Salesianos Impresiones.

Valdés Paz, Juan
1997   *Procesos agrarios en Cuba: 1959–1995.* Havana: Editorial de Ciencias Sociales.

van Veenhuizen, René, ed.
2006   *Cities farming for the future: Urban agriculture for green and productive cities.*
        Ottawa: IDRC.

Varela Pérez, Juan

2008   "Justificado anticipo de zafra." *Granma*, January 2. http://www.granma.cubaweb. cu/2008/01/02/nacional/artic04.html (accessed January 2, 2008).

2009a   "Demandan tareas agrícolas actuales técnicos cada vez más preparados." *Granma*, February 4. http://www.granma.cubaweb.cu/2009/02/04/nacional/artic03. html (accessed February 4, 2009).

2009b   ")Mantendrá el tomate su reinado?" *Granma*, May 26. http://www.granma. cubaweb.cu/2009/05/26/nacional/artic01.html (accessed May 27, 2009).

2009c   "Nada debe limitar a los agricultores urbanos." *Granma*, August 4. http://www. granma.cubaweb.cu/2009/08/04/nacional/artic03.html (accessed August 4, 2009).

2009d   "Si acopio no funciona . . ." *Granma*, February 6. http://www.granma.cubaweb. cu/2009/02/06/nacional/artic02.html (accessed February 6, 2009).

Vázquez, Adelina

2008   "Biofábrica de San José de las Lajas incrementa exportación." *Cadena Habana*, March 21. http://www.cadenahabana.cu/noticias/cientificas/cientificas01210308. htm (accessed May 29, 2009).

Vázquez Moreno, Luis L., Emilio Fernández González, Juan Lauzardo Rico, Tais García Torriente, Janet Alfonso Simonetti, and Rebeca Ramírez Ochoa

2005   "Manejo agroecológico de plagas en fincas de la agricultura urbana (MAPFAU)." Havana: INISAV.

Villavicencio, Daniel

2005   "Country studies: Latin America and the Caribbean." UNESCO. http://portal. unesco.org/education/en/ev.php-URL_ID=54737&URL-sectio=201.html (accessed August 4, 2008).

Vivas, Esther

2008   "Frente a la crisis alimentaria qué alternativa?" *América Latina en Movimiento* 433 (June 24): 23–24.

2009   "Otra agricultura para otro clima." *Rebelión*, March 11. http://www.rebelion.org/ noticia.php?id=44550 (accessed November 4, 2009).

Wagner, Caroline S., Irene Brahmakulam, Brian Jackson, Anny Wong, and Tatsuro Yoda

2001   *Science and technology collaboration: Building capacity in developing countries?* Santa Monica, Calif.: RAND Corporation.

Weis, Tony

2008   *The global food economy: The battle for the future of farming.* London: Zed Press.

Wise, Timothy A.

2010   "The true cost of cheap food." *Resurgence* 259 (March–April). http://www. resurgence.org/magazine/article3035-the-true-cost-of-cheap-food.html (accessed March 10, 2010).

World Development Indicators database

n.d.   "Fertilizer consumption: 100 grams per hectare of arable land—United States (historical data)." Nation Master. http://www.nationmaster.com/time. php?stat=agr_fer_con_100_gra_per_hec_of_ara_lan-granmes-per-hec tare-arable-land&country=us-united-states (accessed July 22, 2009).

Xinhua News Agency

2006 "Centro científico cubano cumple 20 años de exitosa labor." *Xinhuanet,* July 1. http://www.spanish.xinhuanet.com/spanish/2006-07/01/content_276472.htm (accessed April 23, 2009).

Zimbalist, Andrew, Howard J. Sherman, and Stuart Brown

1989 *Comparing economic systems: A political-economic approach.* New York: Harcourt Brace Jovanovich.

Zito, Míriam

2008 "Biotecnología cubana, estrategia para el 2008." Academia de Ciencias de Cuba, January 3. http://www.academiaciencias.cu/paginas/noticia.asp?not=48 (accessed April 23, 2009).

Index

Sinan Koont is associate professor (emeritus) of economics at Dickinson College. He has spent the last ten years researching urban, suburban, and family-based agriculture in Cuba, including doing field work at many sustainable farms on the island.

Contemporary Cuba
Edited by John M. Kirk